Green Energy and Technology

For further volumes:
http://www.springer.com/series/8059

Thomas Bauer

Thermophotovoltaics

Basic Principles and Critical Aspects of System Design

Springer

Dr. Thomas Bauer
Deutsches Zentrum für
Luft- und Raumfahrt e.V. (DLR)
Institut für Technische Thermodynamik
Pfaffenwaldring 38-40
70569 Stuttgart
Germany
e-mail: thomas.bauer@dlr.de

ISSN 1865-3529 e-ISSN 1865-3537

ISBN 978-3-642-19964-6 e-ISBN 978-3-642-19965-3

DOI 10.1007/978-3-642-19965-3

Springer Heidelberg Dordrecht London New York

Cover design: eStudio Calamar, Berlin/Figueres

Printed on acid-free paper

Springer is part of Springer Science+Business Media (www.springer.com)

Preface

Thermophotovoltaic (TPV) is a direct heat to electricity conversion method that was first suggested in the 1950s. The TPV conversion concept is straightforward. Thermal or infrared radiation is converted by a photovoltaic (PV) cell into electricity, just in the same way as solar radiation is converted by a PV cell into electricity. A typical TPV system consists of a high-temperature radiator (1,000–1,700°C), a cavity that includes a filter to control the infrared spectrum and PV cells to convert thermal radiation into electricity. In contrast to solar PV, TPV has two decisive advantages. First, TPV conversion is applicable to any high temperature heat source including the solar, combustion, nuclear and waste heat sources. Second, the efficiency can be enhanced by the control of the absorbed spectrum in the PV cell compared to solar PV. Even though TPV conversion has these advantages, systems are still in a research and development phase at present. At the time of writing, research activities in TPV conversion see a decline. It is the author's hope that this book can contribute to the development of the technology and that the funding situation improves. This would allow demonstration of not only high performance single components, but also systems with high efficiencies. In the past, there have been other hindrances in the development of the technology. For example, there was a lack of suitable high performance PV cells. At the present, the authors view is that the variety of the engineering disciplines involved in an effective TPV design are the major hindrance. The research background of TPV is often solar PV. However, an effective TPV system design requires also the involvement of other disciplines. These essential engineering disciplines include areas such as heat and mass transfer, optics with infrared filters, high temperature materials with ceramics, incandescent lamp with vacuum designs and radiant burners. A particular aim of this book is to emphasise the role of other engineering disciplines and their contribution to the TPV system development as well as material aspects to be considered for the components (filters, radiator, PV cells).

Chapter 1 consists of a general TPV discussion related to the field of electricity generation, the difference between solar PV and TPV, scientific literature sources, the historical development and the definition of conversion efficiencies. The remainder of this book is divided into two parts. The first part is related to three major single components of a TPV system. These are the radiator in Chap. 2, the filter in Chap. 3 and the PV cell in Chap. 4. The second part focuses on systems. Chapter 5 discusses heat transfer aspects in terms of the underlying theory and modelling methods for systems. Chapter 6, the heart of this book, focuses on the cavity design and optical control in the system. Chapter 7 discusses competing technologies and identifies specific TPV strengths. In the assessment in Chap. 8 potential TPV applications are identified.

The material of this book is largely based on a Ph.D. thesis. The author wishes to extend special thanks to his former supervisors Prof. Nicola Pearsall, Dr. Ian Forbes and Dr. Roger Penlington from Northumbria University at Newcastle upon Tyne in the UK.

Stuttgart, May 2011 Thomas Bauer

Contents

Abbreviations

1D, 2D, 3D	One, two or three dimensional
AC	Alternating current
AFC	Alkaline fuel cell
AKS	Compound based on aluminium, potassium (Kalium) and silicon
AlAs	Aluminum arsenide
Al_2O_3	Aluminum oxide, alumina, sapphire
AlN	Aluminum nitride
$Al_{23}O_{27}N_5$	Aluminum oxynitride
AlP	Aluminum phosphide
AlSb	Aluminum antimonide
AM	Air mass
AMTEC	Alkali metal thermal-to-electric converter
APU	Auxiliary power unit
AR	Antireflective
Au	Gold
BASE	Beta alumina solid electrolyte
B_4C	Boron carbide
BaO	Barium oxide
$BaTiO_3$	Barium titanate
BeO	Beryllium oxide, beryllia
B_2O_3	Boron trioxide
BSR	Back surface reflector
BN	Boron nitride
BP	Boron phosphide
C	Graphite
CaO	Calcium oxide
CdTe	Cadmium telluride
Ce_2O_3	Ceria
CeS	Cerium monosulfide
CH_4	Methane
CH_3OH	Methanol

CHP	Combined heat and power
CIO	Cadmium indium oxide, $CdO-In_2O_3$
CO	Carbon monoxide
CO_2	Carbon dioxide
CTO	Cadmium stannate, Cd_2SnO_4
$CuInGaSe_2$	Copper indium gallium selenide
CVD	Chemical vapor deposition
DC	Direct current
DMFC	Direct methanol fuel cell
Er_2O_3	Erbium oxide
EU	European Union
FF	Fill factor
FSF	Front surface filter
FSS	Frequency selective surface
GaAs	Gallium arsenide
GaInP	Gallium indium phosphide
GaSb	Gallium antimonide
GaTe	Gallium telluride
GaP	Gallium phosphide
Ge	Germanium
HfC	Hafnium carbide
HfO_2	Hafnium oxide, Hafnia
Ho_2O_3	Holmium oxide
H_xC_x	Hydrocarbons
ICE	Internal combustion engine
IEEE	Institute of Electrical and Electronics Engineers
InAs	Indium arsenide
InGaAs	Indium gallium arsenide
InGaAsSb	Indium gallium arsenide antimonide, also GaInAsSb
InAsSbP	Indium arsenide antimonide phosphide
In_2O_3	Indium oxide
InP	Indium phosphide
InSb	Indium antimonide
Ioffe	Ioffe Physical Technical Institute, Russia
IR	Infrared
ISE	Fraunhofer Institute for Solar Energy Systems, Germany
ISFH	Institut für Solarenergieforschung, Germany
ISET	Institut für Solare Energieversorgungstechnik, Germany
ITO	Indium tin oxide, $In_2O_3-SnO_2$
K_2O	Potassium oxide
LaF_3	Lanthanum fluoride
Laser	Light amplification by stimulated emission of radiation
LED	Light emitting diode
LPE	Liquid phase epitaxy
LPG	Liquefied propane gas

MBE	Molecular beam epitaxy
MCFC	Molten carbonate fuel cell
MEMS	Microelectromechanical systems
MFI	Multi foil insulation
$MgAl_2O_4$	Spinel
MgF_2	Magnesium fluoride
MgO	Magnesium oxide, magnesia
MIM	Monolithic interconnected module
MIS	Metal insulation semiconductor
MIT	Massachusetts Institute of Technology
MTPV	Micron-gap or microscale TPV, this book uses instead NF-TPV and MEMS-TPV
Mo	Molybdenum
$MoSi_2$	Molybdenum disilicide
MOCVD	Metal organic vapor phase epitaxy, also organometallic vapor phase epitaxy-MOVPE
Na_2O	Sodium oxide
N/A	Not available
NF-TPV	Near-field thermophotovoltaics
NREL	National Renewable Energy Laboratory, US
NPAC	Northumbria Photovoltaics Applications Centre, UK
NO_x	Nitrogen oxides
OC	Open circuit
PAFC	Phosphoric acid fuel cell
PCA	Polycrystalline alumina
PCM	Phase change material
PV	Photovoltaic, photovoltaics
psi	Pound per square inch (1 psi = 6895 Pa)
PSI	Paul Scherrer Institute, Switzerland
Pt	Platinum
ppm	Parts per million
QE	Quantum efficiency
QWC	Quantum well cell
RCF	Refractory ceramic fiber
RTE	Radiative transfer equation
S	Sulfur
Sb_2S_3	Antimony sulfide
Sb_2Se_3	Antimony selenide
SC	Short circuit
SERC	Solar Energy Research Center, Sweden
SEC	Specific energy consumption
Si	Silicon
SiC	Silicon carbide
SiGe	Silicon germanium
SiO_2	Fused silica (quartz glass)

Si_3N_4	Silicon nitride
SnO_2	Tin oxide
SOFC	Solid oxide fuel cell
SO_x	Sulfur oxides
SPFC	Solid polymer membrane fuel cell
$SrTiO_3$	Strontium titanate
STPV	Solar thermophotovoltaic
Ta	Tantalum
$TaSi_2$	Tantalum silicide
TCO	Transparent conducting oxide
TE	Thermoelectric, thermoelectrics
ThO_2	Thorium dioxide, Thoria
TiB_2	Titanium diboride
TiO_2	Titanium (IV) oxide
TPX	Thermophotonic, thermophotonics
TPV	Thermophotovoltaic, thermophotovoltaics
UE	Ultimate efficiency
UNSW	University of New South Wales, Australia
UPS	Uninterruptible power supply
W	Tungsten
WC	Tungsten carbide
YAG	Yttrium aluminum garnet, $Y_3A_{15}O_{12}$
YF_3	Yttrium fluoride
Y_2O_3	Yttrium oxide, yttria
Yb_2O_3	Ytterbium oxide, ytterbia
ZnO	Zinc oxide
ZnS	Zinc sulfide
ZnSe	Zinc selenide
ZrC	Zirconium carbide
ZrO_2	Zirconium oxide, zirconia

Nomenclature

A	Area, cross section (m^2)
A	Thermionic constant ($A\ K^{-2}\ m^{-2}$)
A_a	Aperture area of a solar concentrator (m^2)
A_r	Receiver area of a solar concentrator (m^2)
b	Constant of Wien's displacement law $b = 2.898 \cdot 10^{-3}$mK
c	Speed of radiation in a medium with refractive index n ($m\ s^{-1}$)
c_0	Speed of radiation in vacuum $c_0 = 2.99792 \cdot 108\ m\ s^{-1}$
d	Diameter (m)
D_e	Electron diffusion coefficient ($m^2\ s^{-1}$)
D_h	Hole diffusion coefficient ($m^2\ s^{-1}$)
e_0	Elementary charge $e_0 = 1.60218 \cdot 10^{-19}$ A s
f	Sticking coefficient of the Hertz–Langmuir equation
F	Factor for the molar melting enthalpy (l)
F_{h-c}	View factor from the hot to the cold plate (l)
h	Planck's constant h = $6.62617 \cdot 10^{-34}$ J s
h	Heat-transfer coefficient for convection heat transfer ($W\ m^{-2}\ K^{-1}$)
h_v	Photon energy (J)
h_{vg}	Bandgap energy (J)
H	Heat flux ($W\ m^{-2}$)
$H_{m,\ mol}$	Molar melting enthalpy (J/mol)
i_b	Total blackbody intensity ($W\ m^{-2}$)
$i_{b\lambda}$	Planck's function depending on wavelength ($W\ m^{-2}\ \mu m^{-1}$)
i_{bv}	Planck's function depending on frequency ($W\ m^{-2}$ s)
$I_{(v)}$	Radiation intensity depending on frequency ($W\ m^{-2}$)
I_b	Total blackbody radiation intensity ($W\ m^{-2}$)
$I_{b\lambda}$	Blackbody radiation intensity depending on wavelength ($W\ m^{-2}\ \mu m^{-1}$)
I_{bv}	Blackbody radiation intensity depending on frequency ($W\ m^{-2}$ s)
I	Radiation intensity ($W\ m^{-2}\ \mu m^{-1}$)
$I_{0,\ \lambda}$	Initial radiation intensity ($W\ m^{-2}$)
$I_{S,\ \lambda}$	Radiation intensity after path length S ($W\ m^{-2}$)
I_T	Blackbody radiation for ultimate band model-TPV case ($W\ m^{-2}$)

J	Current density of the PV cell (A m^{-2})
J_m	Current density at the maximum power point of the PV cell (A m^{-2})
J_{max}	Maximum photocurrent density (A m^{-2})
J_{ph}	Photocurrent density (A m^{-2})
J_s	Dark saturation current density (A m^{-2})
k	Thermal conductivity (W m^{-1} K^{-1})
k	Boltzmann constant $k = 1.38066 \cdot 10^{-23}$ J K^{-1}
k_R	Radiative conductivity of the Rosseland approximation (W m^{-1} K^{-1})
$k_{(\lambda)}$	Extinction coefficient (1)
L	Thickness (m)
L_e	Electron diffusion length (m)
L_h	Hole diffusion length (m)
L_p	Mean free molecular path (cm)
M	Molecular weight (g/mol)
m	Mass (kg)
m_e	Electron mass $m_e = 9.1094 \cdot 10^{-31}$ kg
m^*	Effective mass (kg)
n	Refractive index (1)
n_i	Intrinsic concentration (m^{-3})
n_{ph}	Number of photons (1)
N	Number, number of charge carriers (1)
N_A	Acceptor doping concentrations (m^{-3})
N_D	Donor doping concentrations (m^{-3})
p_v	Vapor pressure (torr)
p	Pressure (torr)
P	Power (W), power density (W m^{-2}), heat transfer rate (W)
P_{el}	Electrical output power of PV cell at maximum power point (W)
$P_{el,\ net}$	P_{el} minus house keeping power (W)
$P_{el,\ meas}$	Measured electrical power density of the PV cell (W m^{-2})
P_{heat}	Heat output of the PV cell (W)
P_{input}	Total input power of the TPV system (W)
P_{net}	Net heat transfer to the radiator (W)
P_{PV}	Net absorbed power of PV cell with conduction/convection (W)
$P_{PV\#}$	Net absorbed radiative power of the PV cell (W)
P_{solar}	Ultimate power density limit of solar PV conversion (W m^{-2})
P_{TPV}	Ultimate power density limit of TPV conversion (W m^{-2})
Q	Quality factor of the transparent conduction oxide filter
Q_s	Photon number with energies greater $h\nu_g$-solar PV case (m^{-2} s^{-1})
Q_T	Photon number with energies greater $h\nu_g$ - TPV case (m^{-2} s^{-1})
r_s	Sun radius $r_s = 6.96 \cdot 10^8$ m
r_{se}	Sun-earth distance $r_{se} = 1.50 \cdot 10^{11}$ m
r_{large}	Large radius for conduction heat transfer (m)
r_{small}	Small radius for conduction heat transfer (m)
$r+$	Ratio of upper spectral band divided by total radiation above x_g (1)
$r-$	Ratio of lower spectral band divided by total radiation below x_g (1)

R	Molar gas constant $R = 8.314472$ J/(mol · K)
R_L	Electrical resistance of the load (Ω)
$R_{(v)}$	Reflectivity depending on the radiation frequency (l)
R_{th}	Thermal resistance (K/W)
$\vec{S}$	Direction vector for the radiative transfer equation (l)
S	Path length (m)
S	Seebeck coefficient (V K^{-1})
S	Shape factor for conduction heat transfer (m)
$SR_{(v)}$	Spectral response (A W^{-1})
T	Temperature (K)
T_a	Absorber temperature (K)
T_{AV}	Average temperature (K)
T_c	Cold side temperature (K)
T_{cell}	PV cell temperature (K)
T_F	Fluid temperature for convection heat transfer (K)
T_h	Hot side temperature (K)
T_m	Melting temperature (K)
T_s	Source temperature, sun surface temperature (K)
T_w	Wall temperature for convection heat transfer (K)
v	Frequency (s^{-1})
$v-$	Frequency band limit below v_g (s^{-1})
$v+$	Frequency band limit above v_g (s^{-1})
v_g	Bandgap frequency (s^{-1})
v_p	Plasma frequency (s^{-1})
V	Voltage (V)
V_c	Thermal voltage $V_c = (kT_{cell})/e_0$ (V)
V_g	Bandgap voltage $V_g = (hv_g)/e_0$ (V)
V_m	Voltage at the maximum power point of the PV cell (V)
V_{oc}	Open circuits voltage (V)
W	Evaporation rate (g cm^{-2} s^{-1})
x	Substitution $x = (hv)/(kT_s)$ (l)
$\tilde{x}$	Substitution $\tilde{x} = (hc_0)/(kT_s n\lambda)$ (l)
x_g	Normalized bandgap $x_g = (hv_g)/(kT_s)$ (l)
$x-$	Band limit below x_g (l)
$x+$	Band limit above x_g (l)
Y	Relaxation frequency of the semiconductor (s^{-1})
Z	Figure of merit for thermoelectric generator (K^{-1})

α	Absorption coefficient (m^{-1})
β	Parameter for the dark saturation current density (A cm^{-2})
ε	Emissivity (l)
ε_0	Dielectric constant in vacuum = $8.85418 \cdot 10^{-14}$ F m^{-1}
ε_b	Dielectric constant of the semiconductor (l)
η	Efficiency (l)

η_{cavity}	Efficiency of the cavity (l)
η_{source}	Efficiency of the heat source (l)
$\eta_{TE, max}$	Maximum efficiency for thermoelectric generators (l)
η_{PV}	Efficiency of the PV cell with conduction and convection (l)
η_{TPV}	Electrical thermophotovoltaic efficiency (l)
$\eta_{TPV, CHP}$	Combined heat and power thermophotovoltaic efficiency (l)
η_{house}	Efficiency of house keeping power sources (l)
η_{sys}	Electrical efficiency of the TPV system (l)
$\eta_{sys, CHP}$	Combined heat and power efficiency of the TPV system (l)
η_{Carnot}	Carnot efficiency (l)
η_{OC}	Voltage factor (l)
$\eta_{OC, FF, QE}$	Ideal PV cell related efficiency (l)
η_{QE}	Collection efficiency (l)
$\eta_{QE, ext(\nu)}$	External quantum efficiency depending on the frequency (l)
$\eta_{QE, int(\nu)}$	Internal quantum efficiency depending on the frequency (l)
η_{FF}	Fill factor (l)
η_{UE}	Ultimate efficiency (l)
$\eta_{UE, TPV}$	Ultimate efficiency of TPV (l)
$\eta_{UE, solar}$	Ultimate efficiency of solar PV (l)
η_{Array}	PV cell array efficiency (l)
θ	Angle, zenith angle (°)
θ_s	Opening half angle of the sun at the Earth's distance (°)
λ	Wavelength (μm)
λ_{max}	Wavelength where Planck's function becomes maximum (μm)
μ	Electron mobility of the semiconductor ($m^2 V^{-1} s^{-1}$)
π	Constant $\pi = 3.14159$ (l)
ρ	Reflection (l)
$\rho\perp$	Reflection, polarization perpendicular to plane of incidence (l)
ρ_{II}	Reflection, polarization in plane of incidence (l)
σ	Stefan-Boltzmann constant $\sigma = 5.670 \cdot 10^{-8}$ W m^{-2} K^{-4}
σ	Electrical conductivity (S m^{-1})
Φ	Work function (J)

Chapter 1
Introduction

1.1 Importance of Thermophotovoltaics

1.1.1 Energy Aspect

In modern industrial society, the majority of energy is consumed in the sectors of transportation, building and industry and fossil fuels are the primary sources of this energy. The use of fossil fuels has led to worldwide concerns about security of supply, increasing energy demand, limitation of resources and local and global environment impacts (e.g. acid rain and climate change). A consequence is increased interest in non-fossil fuel energy resources and the efficient use of fossil fuels. Thermophotovoltaics (TPV) converts heat directly into electricity and has been examined in all major energy sectors for both non-fossil fuel energy resources (e.g. radioactive, solar heat and biomass) and the efficient use of fossil fuel. Potentially, TPV systems can convert heat into electricity with Carnot efficiency, which would make it an attractive alternative to existing electricity generation technologies. At the current stage of research high efficiencies have not been demonstrated and it is uncertain which practical efficiencies TPV systems can achieve. However, even moderate (partly) demonstrated efficiencies make TPV conversion already attractive for the efficient use of fossil fuels in applications such as combined heat and power (CHP), portable power and waste heat recovery. Currently, TPV systems are mainly developed for fossil fuel powered combustion applications, which are not favourable from the energy saving aspect. However, the fuel flexibility of TPV systems allows the change from fossil fuels to bio-fuels in future, which may be more difficult for other technologies (e.g. fuel cells). Hence, TPV conversion could solve some of the fossil fuel constraints, if market and technology challenges can be overcomed.

T. Bauer, *Thermophotovoltaics*, Green Energy and Technology,

DOI: 10.1007/978-3-642-19965-3_1,

1.1.2 Technology Aspect

TPV conversion has inherently some technological properties which makes it advantageous compared to existing routes to supply electricity. At the moment, the two major routes are the large-scale grid connected supply and the small-scale battery market.

In industrial countries the vast amount of electricity is centrally generated in large power plants using internal and external heat engines (e.g. gas and steam turbines) together with a generator, and transmitted and distributed via the grid. One major disadvantages of central power generation is the waste of large amounts of heat. Even modern fossil fuel powered combined cycle power plants discharge about half of their input as waste heat. Other disadvantages include the system complexity, security of supply issues (e.g. black outs) and distribution and transmission losses. Hence, a rather distributed generation of heat and electricity would be desirable. However, downscaling of central power technologies is critical in terms of the efficiency reduction, higher maintenance and high noise. Renewable generators (e.g. wind and solar photovoltaics) can replace large-scale fossil power plants. However, their widespread use requires economic high power and high capacity electrical storage systems that have been not identified yet. Another possibility is the utilisation of more steady renewable sources in more remote locations. Examples include offshore wind parks and solar thermal power plants. The latter one offers the advantage of an affordable thermal energy storage system that can operate the plant for several hours. Large-scale utilisation of technologies such as offshore wind and solar thermal power will require major investments in the electrical grid (e.g. high voltage direct current power lines).

On the small-scale power range, in the order of milliwatts to hundreds of watts, a large amount of non-grid connected electricity originates from (rechargeable) batteries, which have disadvantages in terms of a limited lifetime, a slow charging process and a low gravimetric energy density (MJ/kg). Hydrocarbon fuels have in the order of 100 times higher gravimetric energy densities and can be easily stored and quickly supplied [1]. Even a low efficient TPV system could have superior properties compared to batteries. Hence, TPV conversion is a promising technology to convert hydrocarbon fuels into electricity. In this way, a lightweight and quickly rechargeable portable power generator could be realised.

It can be concluded that both, batteries and large power plants using the grid have their own disadvantages to supply electricity. TPV conversion is one of several other technologies in a research and development stage that offer advantages over this existing electricity supply infrastructure. This is especially true for the intermediate power range (around 10 W to 10 kW). The advantages include high reliability, low noise, high gravimetric and volumetric energy density, portability and long operation time. Hence, TPV has the potential to replace some of the existing grid and battery infrastructure in the intermediate power range in future.

1.2 Comparison of Solar PV and TPV Conversion

Both solar photovoltaics (PV) and thermophotovoltaics (TPV) use photovoltaic cells to generate electricity from the radiation of a high-temperature thermal source. The key differences are the geometry and the heat source temperature [2].

1.2.1 Solar Photovoltaics

The sun surface radiation corresponds approximately to 5,800 K blackbody radiation. The intensity (W/m^2) drops over the large distance to a maximum value of approximately 0.1 W/cm^2 at the earth's surface (Fig. 1.1 upper part) [3, 4]. A major part of the solar radiation is in the visible spectral range according to Planck's radiation law. Solar PV operates rather unsteadily in terms of intensity, spectrum and angle of solar radiation because these parameters depend on location (latitude, longitude), different cycles (solar, annual, seasonal, daily), cloud cover and atmospheric absorption. Another variable parameter of solar PV is the PV cell temperature. For non-concentrator systems, cell temperature is usually not controlled and depends on the environment. For low temperatures, the cell efficiency increases and vice versa the efficiency decreases for high temperatures. The complexity of a cooling system for the hot cells is usually not economically worthwhile.

For solar PV, the radiation intensity is the boundary condition with a given magnitude and spectrum. Scattering and absorption of the solar radiation by the earth's atmosphere is taken into account by different standardised spectra (AM0, AM1, AM1.5) [5]. Assuming a solar spectrum, the ideal PV cell bandgap can be identified. It is important to note that, in contrast to TPV conversion, an ideal bandgap maximises both the efficiency and the electrical power density. Assuming a typical solar intensity of 0.1 W/cm^2 and a conversion efficiency of 20% leads to a maximum electrical power density of about 0.02 W/cm^2. For non-concentrator solar PV, CHP operation is usually not utilised, so that the remaining heat is lost (for the assumption 0.08 W/cm^2).

1.2.2 Thermophotovoltaics

TPV conversion, on the other hand, can use a variety of heat sources, which usually heat up a radiator, also named emitter, to typically temperatures T_s of 1,300–2,000 K (Fig. 1.1 lower part). There are a few arrangements imaginable where no radiator may be required including the direct conversion of flame radiation and the conversion of waste heat where the process already radiates with a suitable spectrum. This temperature range leads to a theoretical hemispherical

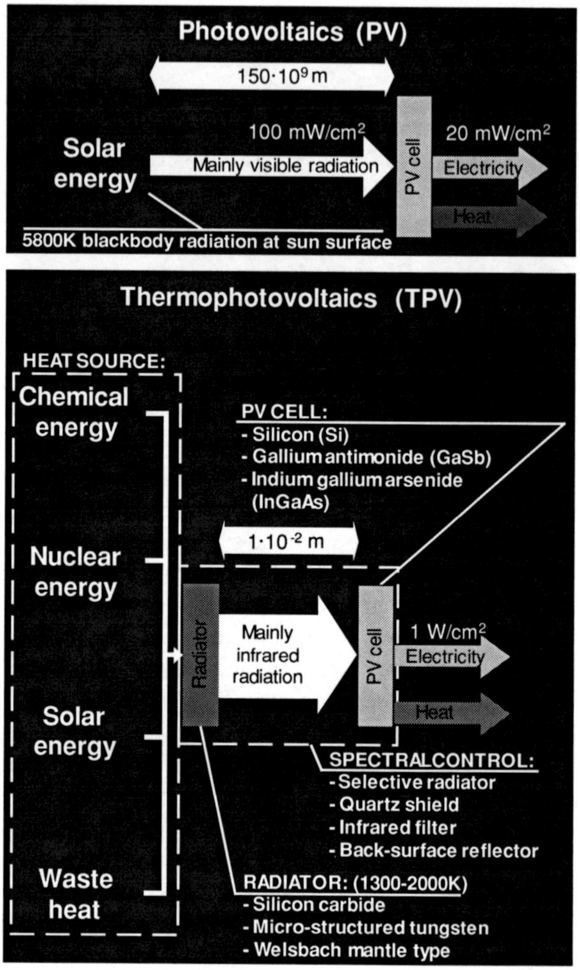

Fig. 1.1 Schematic to compare non-concentrator solar PV and TPV conversion. The figure shows potential heat sources and the major components with examples of a TPV system

total radiation per unit area of approximately 16–91 W/cm^2 according to the Stefan-Boltzmann law with an emissivity of one (σT_s^4). The major part of this radiation is in the infrared spectral range according to Planck's radiation law. Ideally no radiation is lost due to the close arrangement of radiator and PV cells. This has allowed demonstration of high electrical power densities of the PV cell with values over 2.5 W/cm^2. TPV systems typically operate steadily in terms of intensity, spectrum and angle of radiation as well as PV cell temperature. For example in waste heat recovery applications these systems could operate steadily 24 h a day and 365 days per year, where operation parameters may only alter due to some wear and ageing of the TPV system. Defining TPV efficiency is more complex compared to solar PV. The total input may be defined for example as the calorific value of a fuel or as a heat flux. The high power densities in TPV conversion make CHP operation possible, so that the useful output can be either purely electricity or electricity and heat. Therefore, the definition of TPV efficiency as useful output to total input is ambiguous.

TPV system design usually requires a spectral control concept, as well as selection and design of the heat source, radiator and PV cell each with their own limitations and qualities. Figure 1.1 shows some examples of the various concepts and component options. Spectral control is a method to spectrally match the absorbed power of the PV cell in accordance to the cell bandgap and often includes additional components such as filters. Alternatively, the spectrum can be spectrally controlled by a selective radiator and by filters and mirrors within the PV cell. In the solar PV case, photons with energies below the PV cell bandgap energy (out-of-band radiation) are lost. TPV systems, however, can suppress or recover these photons by some form of spectral control to increase the efficiency. This spectral control option potentially leads to higher efficiency of TPV conversion compared to solar PV conversion. Although increased TPV efficiency has been only partly demonstrated on a component but not on the system level. The CHP operation mode has been demonstrated on a system level and the high overall efficiency can be of interest in applications such as for domestic micro CHP systems.

It can be summarised that TPV system design is more complex, but has advantages including steady operation, potentially higher efficiency and heat source flexibility, if compared to solar PV. Additionally, TPV conversion offers effective CHP operation and has demonstrated high electrical power densities compared to non-concentrator solar PV.

1.3 TPV Literature

Books and book chapters are useful sources to introduce TPV technology. In particular these include one chapter [6] and one section [7] by Coutts. A book by Chubb includes detailed theoretical background on the aspects of radiative heat transfer and filters [8]. Other book chapters focus on the solar thermophotovotaic (STPV) conversion, such as work by Fahrenbruch and Bube [9], Green [10] and Würfel [4]. Additionally, sections in direct energy conversion books briefly discuss TPV conversion. Examples are a book by Decher [11] and an older book by Angrist [12].

In 1995, Broman published a bibliography of TPV, which contains 180 entries on TPV conversion of energy for the time period from 1950 to 1994 [13]. A comprehensive TPV review article is available from Coutts, which was published in 1999 [2]. Basu et al. published a review article on microscale radiation principles in the design of different components [14].

Currently the Conference on Thermophotovoltaic Generation of Electricity is the major international event on TPV research. The first four conferences were held under the auspices of the National Renewable Energy Laboratory (NREL) in Colorado, US. Some further conferences were held in Europe. The fifth conference was held in Rome in 2002, the sixth conference in Freiburg/Germany in 2004, the seventh in Madrid in 2007 and the eight in the Palm Desert in 2008. Other conferences with major TPV contributions include the European Photovoltaic Solar Energy Conference, the IEEE Photovoltaic Specialists Conference and the Intersociety Energy Conversion Engineering Conference.

TPV articles can be found in various photovoltaics, physics, materials and energy journals. In 2003, the journal Semiconductor Science and Technology released a special issue about TPV edited by Barnham, Connolly and Rohr [15]. At the time of writing more than thousand publications, such as journal and conference papers, as well as reports have been published. Worldwide, more than hundred patents have been filed on TPV components or systems. The number of PhD and Master thesis account for at least several tens worldwide. In spite of the large number of publications, there is limited literature on the basic principles, critical aspects of system design and the lessons learned from previous developments.

1.4 Historical Development

The following paragraph gives a brief introduction in the historical development of TPV conversion for electricity generation. Other authors give a more detailed discussion. They include Fraas [16], Coutts [2], Ralph and FitzGerald [17], Noreen and Honghua [18] and Nelson [1, 19].

The invention of TPV dates back to about 1956. Most literature references cite Aigrain as the inventor of TPV, who proposed the concept during a series of lectures at the MIT in 1956 [2, 20]. Nelson reported about a demonstration TPV system at the MIT Lincoln Laboratory by Kolm and a publication titled "Solar-battery power source" in the same year [19, 21]. Early US research until the mid 1970s focussed on standalone military power generators with low noise using fossil fuel combustion as a heat source. In this early period the three principal heat sources (solar, nuclear and combustion) and spectral control options (selective radiator, filter, PV cell front and back surface reflector) had been identified [19, 22–24]. Of the available PV cell materials, both silicon and germanium had their own difficulties. The high silicon bandgap usually requires high radiator temperatures, where engineering difficulties occur. Germanium PV cells require lower radiator temperature, but these cells had a low performance [2]. In the mid 1970s, the pace of TPV development slowed significantly when the US Army chose thermoelectric technology as a power source and General Motors discontinued TPV research, because of business challenges during the oil crises [19].

However, overall TPV research profited from the research effort after the energy crisis. Solar energy had been identified as one important non-fossil fuel source. Advances were made in solar TPV (STPV) in the US and Europe, and TPV technology has also profited from progress in PV cell developments. Improvements were made in PV cells using III–V materials and PV concentrator systems. Examples are III–V PV cells, such as gallium antimonide (GaSb) and indium gallium arsenide (InGaAs), which are now commonly used in TPV systems [19].

In addition experience from PV concentrator systems with high radiation densities could be applied to TPV systems (e.g. PV cell design and cooling). TPV research has also benefited from research in the efficient use of fossil fuel and in particular in the area of efficient and durable radiant burners for heating purposes. For example, fibrous rare earth oxide burners [25] and ceramic radiant tube burners [26] have been initially developed for other purposes. These burners have been applied in combustion-powered TPV systems to efficiently convert fuel into radiation.

In the early 1990s there had been a renewed TPV interest for military and space power in the US. At this time two spin-off companies were founded to produce Zn-diffused GaSb cells with a license from Boing (JX-Crystals Inc. and EDTEK Inc.). JX-Crystals developed a propane powered CHP stove (Midnight Sun®) of which 20 units were sold for beta testing from 1998 to 2001. The unit has a fuel input power of 7.5 kW and electrical output power of 150 W [27, 28]. TPV power generators using radioisotope heat sources were considered for deep space missions, where solar PV is not attractive because of the low light level. The Vehicle Research Institute at Western Washington University demonstrated successfully a combustion-TPV generator in a hybrid car [29, 30].

Industrial waste heat recovery using TPV conversion is another promising application and research area, which was proposed by Coutts [2] in the late 1990s (see Sect. 8.5). Also, in the late 1990s fundamental research in the area of Near-field TPV (NF-TPV) started. In NF-TPV the distance between the radiator and the cell is reduced to sub-wavelength dimensions. Classical radiative heat transfer laws cannot describe the radiative heat transfer between radiator and cell. Microscale radiation theory predicts that the radiative heat transfer can be enhanced significantly (see Sect. 6.5.2).

From the early 2000s the development of miniaturised TPV generators with an electrical power below 10 W has accelerated. Microelectromechanical systems or MEMS is a growing research area and aims for the development of micro-scale devices with mechanical elements, sensors, actuators, and electronics using micro-fabrication technology. With this miniaturisation there is a demand to replace the low energy density batteries by an alternative conversion technology using hydrogen or hydrocarbons as a fuel. Among other candidates (e.g. fuel cells, micro turbine), TPV generators are also examined. Some of these TPV generators are based on micro-fabrication techniques.

1.5 Energy Balance and Efficiency of a General Type System

The efficiency of a TPV system depends generally on the definition of the boundaries. The discussion follows Fig. 1.2, which shows schematically the energy flow in a general type of TPV system.

Efficiency is generally defined as the ratio of the useful output to the total input. The useful output for a TPV system may be either the generated electrical power

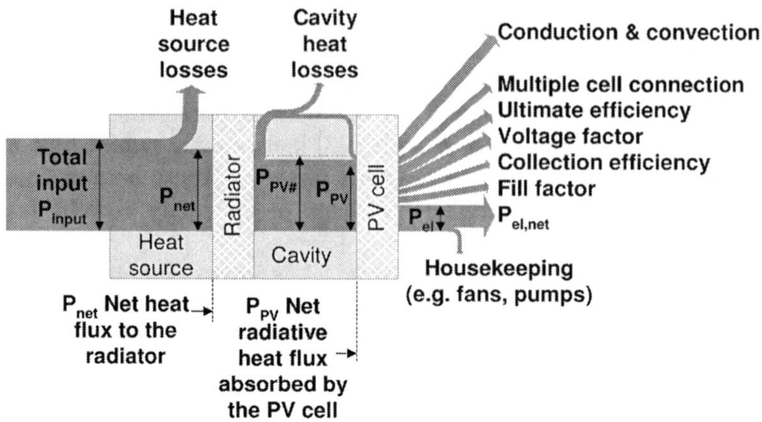

Fig. 1.2 Sankey diagram of a general type of TPV system. The thickness of each energy flow path is not proportional in size

or the generated power and the useful heat for CHP operation. This makes a TPV system using purely electricity inherently less efficient than a CHP system. Furthermore the output power may or may not include power necessary for system housekeeping (for example fuel pump or electric control). The total input depends on the heat source and can be defined in different ways. Four heat sources can be distinguished, namely chemical energy (e.g. hydrocarbon combustion), nuclear energy (e.g. radioisotope heat), solar energy and waste heat (e.g. from industrial high-temperature processes). In a radioisotope, solar or waste heat recovery TPV system the total input may be defined as a heat flux (W/m^2), where heat source losses can often be small. For solar TPV, the optical efficiency of the concentrator can be defined as the heat source efficiency. In a hydrocarbon combustion system the total input may be defined by the fuel flow rate (e.g. kilogram per hour). This rate can be converted into an input power (W) assuming gross or net calorific values of the fuel, where some values in megajoules per kilogram are given in Sect. 8.4.1. For example, efficient high-temperature radiant burners that operate with natural gas can have heat losses of about 20% [26]. This heat is mostly lost through the hot flue gas. The percentage value suggests that high efficiency hydrocarbon powered TPV systems can be configured. On the other hand it can also be pointed out that radioisotope and waste heat recovery TPV systems are inherently more efficient compared to combustion systems. The heat source efficiency can be defined as the ratio of radiator net heat transfer to the total input (e.g. product of fuel flow rate and calorific value), as also shown in Fig. 1.2 Eq. (1.1).

$$\eta_{\mathrm{source}} = \frac{P_{\mathrm{net}}}{P_{\mathrm{input}}} \tag{1.1}$$

Heat transfer in the cavity is dominated by radiation. Convection and conduction heat transfer is parasitic and needs to be minimised. Components in the cavity, such as heat shields, mirrors and the PV cell absorb, reflect and reemit radiation. In steady-state there will be a net heat flow and this flow is considered in Fig. 1.2. A net radiative heat flux absorbed by the PV cell P_{PV} includes short wavelength radiation that can be converted by the PV cell (in-band radiation) as well as long wavelength radiation that cannot be converted by the PV cell (out-of-band radiation).

The PV cell can absorb additional heat by parasitic convection and conduction. The total net heat flux absorbed by the PV cell $P_{PV\#}$ contains both, the radiative and the parasitic conductive/convective heat flux (Fig. 1.2). Two general types of parasitic heat transfer paths to the PV cell can be distinguished. First, parasitic heat transfer can occur directly from the radiator via the cavity gas (e.g. air or inert gas) to the PV cell. The heat transfer modes are gas conduction and possibly free convection. Heat transfer via this path may be minimised by the use of a suitable gas or reduced pressure (vacuum) in the cavity. Another parasitic heat transfer path from the radiator to the PV cell forms usually the (reflective) insulation. Aspects such as parasitic radiation absorption of mirrors can contribute to higher losses via this path. Finally, TPV systems have losses from the cavity via the insulation to the surroundings (cavity heat losses in Fig. 1.2). In particular small sized systems are susceptible to losses to the environment due to the large surface-to-volume ratio.

The cavity efficiency can be defined as the ratio of net radiative absorbed heat flux by the PV cell P_{PV} to net heat transfer to the radiator P_{net} Eq. (1.2). This cavity efficiency is the product of two partial cavity efficiencies, which consider the losses through the insulation to the surroundings and the losses by parasitic direct and indirect conductive/convective heat transfer from the radiator to the cell Eq. (1.2).

$$\eta_{cavity} = \underbrace{\frac{P_{PV\#}}{P_{net}}}_{\text{insulation}} \cdot \underbrace{\frac{P_{PV}}{P_{PV\#}}}_{\substack{\text{conduction}\\ \text{\& convection}}} = \frac{P_{PV}}{P_{net}} \tag{1.2}$$

In some cases it may be not feasible to separate between heat source and cavity losses, as shown in Fig. 1.2, because the heat source with the combustion products and the cavity are spatially not separated (e.g. for hydrocarbon combustion using a Welsbach mantle radiator).

The PV cell efficiency can be defined as the ratio of electricity output P_{el} to the net absorbed radiative heat flux P_{PV} Eq. (1.3). This efficiency depends on several operational conditions including the PV cell temperature, the radiation intensity, as well as the spectral, angular and spatial radiation distribution. So far these operation conditions have not been standardised in order to characterise PV cells for TPV systems. Therefore any PV cell efficiency has to be seen in conjunction with the operation conditions in the TPV system. For example, the efficiency of a

Table 1.1 Examples of demonstrated and projected TPV-system performance

System: Radiator/cell type	Yb_2O_3/Si [31]	W/GaSb [28, 32, 33]	SiC/GaSb MidnightSun® [27, 28]	W/InGaAs [34]
Heat source	Butane combustion	Propane or JP8 combustion	Propane combustion	Radioisotope modules
Input power	2.0 kW	4.4 kW	5.0 kW	0.5 kW
Radiator material	Yb_2O_3 Welsbach mantle	AR coated tungsten foil on SiC	SiC radiator	Tungsten
Radiator temperature	1,462°C	1,275°C	1,200°C	1,077°C
PV cells type	Si (Red enhanced)	GaSb	GaSb	0.6 eV InGaAs MIM
Cell producer	UNSW	JX-Crystal	JX-Crystal	Emcore
Cell temperature	14°C	25°C	75°C	50°C
Cooling: PV cell by	Forced water	Forced water	Forced air	Radiation
Cell power	48 W	700 W	100 W (net 80 W)	100 W
Cell area	481 cm^2	470 cm^2	250 cm^2	200 cm^2
Power density	0.1 W/cm^2	1.5 W/cm^2	0.4 W/cm^2	0.5 W/cm^2
Means of effective cavity design and spectral control	Selective radiator Quartz glass tube Gold plated glass reflector	Selective AR coated tungsten radiator Quartz glass tube Dielectric filter on top of the cell Inert gas filling	Dielectric filter Glass tube	Dielectric-TCO filter on PV cell Optionally BSR Multifoil insulation of the cavity Vacuum in space
η_{source}	Low, no recuperator	75% (measured)	Low, no recup.	100%
$\eta_{source} \times \eta_{TPV}$	2.4% (measured)	16% (projected)	2.0% (measured)	20% (projected)
$\eta_{source} \times \eta_{TPV} \times \eta_{house}$	N/A	N/A	1.6% (measured)	N/A

specific PV cell can be expected to be high for a spectrally matched radiator and cell, but low for the spectrally unmatched case.

$$\eta_{PV} = \frac{P_{el}}{P_{PV}} \tag{1.3}$$

In practice the total net absorbed heat flux of the PV cell can then be determined by measuring the total output of electricity P_{el} and heat P_{heat} of the PV cell Eq. (1.4). It needs to be considered that this PV cell efficiency includes parasitic conductive/convective components and these components depend on the cavity design. Hence, PV cells can be characterised thermally isolated in vacuum in order to minimise these parasitic heat transfer.

$$\eta_{PV\#} = \frac{P_{el}}{P_{PV\#}} = \frac{P_{el}}{P_{heat} + P_{el}} \tag{1.4}$$

The reduction due to housekeeping power is expressed in terms of the house keeping efficiency Eq. (1.5).

$$\eta_{house} = \frac{P_{el,net}}{P_{el}} \tag{1.5}$$

The overall efficiency of a TPV system is then a product of all partial efficiencies. Hence, high system efficiencies require maximisation of *all* partial efficiencies Eq. (1.6).

$$\eta_{sys} = \eta_{source} \cdot \eta_{cavity} \cdot \eta_{PV} \cdot \eta_{house} = \eta_{source} \cdot \eta_{TPV} \cdot \eta_{house} = \frac{P_{el,net}}{P_{input}} \tag{1.6}$$

Table 1.1 gives an overview in terms of the performance of three combustion systems and one radioisotope TPV system. The direct comparison of system efficiency can be misleading. Aspects including the following can have an impact on the system efficiency: type of heat source, system size, radiator operation temperature and PV cell temperature. It also needs to be considered that the electrical power density inherently increases with the radiator temperature.

References

1. Nelson R (2003) TPV Systems and state-of-the-art development. In: Proceedings of the 5th conference on thermophotovoltaic generation of electricity, American Institute of Physics, Rome, pp 3–17, 16–19 Sep 2002
2. Coutts TJ (1999) A review of progress in thermophotovoltaic generation of electricity. Renew Sustain Energy Rev 3(2–3):77–184
3. Sze SM (1981) Physics of semiconductor devices, 2nd edn. Wiley, New York
4. Würfel P (1995) Physik der Solarzellen (in German), 2nd edn. Spektrum Akademischer Verlag, Heidelberg
5. Partain LD (1995) Solar cells and their applications. Wiley Interscience, New York

6. Coutts TJ (2001) Thermophotovoltaic generation of electricity. In: Archer MD, Hill R (eds) Clean electricity from photovoltaics, Chap 11, vol 1. Series on photoconversion of solar energyImperial College Press, London
7. Benner JP, Coutts TJ (2000) Thermophotovoltaics. In: Dorf RC (ed) The electrical engineering handbook. CRC Press, Boca Raton
8. Chubb D (2007) Fundamentals of thermophotovoltaic energy conversion. Elsevier Science, Amsterdam
9. Fahrenbruch AL, Bube RH (1983) Concentrators, concentrator systems, and photoelectrochemical cells. In: Fahrenbruch AL, Bube RH (eds) Fundamentals of solar cells, Chap 12. Aademic Press, Orlando, pp 505–540
10. Green MA (2003) Thermophotovoltaic and thermophotonic conversion. In: Green MA (ed) Third generation photovoltaics—advanced solar energy conversion, Chap 9. Springer, Berlin, pp 112–123
11. Decher R (1997) Direct energy conversion fundamentals of electric power production. Oxford University Press, London
12. Angrist SW (1976) Direct energy conversion, 3rd edn. Allyn and Bacon, Newton, MA
13. Broman L (1995) Thermophotovoltaics bibliography. Prog Photovolt Res Appl 3(1):65–74
14. Basu S, Chen Y-B, Zhang ZM (2007) Microscale radiation in thermophotovoltaic devices—A review. Int J Energy Res 31(6–7):689–716
15. Barnham K, Connolly J, Rohr K (2003) Special issue on thermophotovaltaics (TPV), complete volume with 19 papers. Semiconductor science and technology 18:5 http://iopscience.iop.org/0268-1242/18/5
16. Fraas L, Minkin L (2007) TPV History from 1990 to Present & Future Trends. In: Proceedings of the 7th world conference on thermophotovoltaic generation of electricity, Madrid, 25–27 Sept 2006, American Institute of Physics, pp 17–23
17. Ralph EL, FitzGerald MC (1995) Systems/marketing challenges for TPV. In: Proceedings of the 1st NREL conference on thermophotovoltaic generation of electricity, Copper Mountain, Colorado, US, 24–28 July 1994, American Institute of Physics, pp 315–321
18. Noreen DL, Honghua D (1995) High power density thermophotovoltaic energy conversion. In: Proceedings of the 1st NREL conference on thermophotovoltaic generation of electricity. Copper Mountain, Colorado, US, 24–28 July 1994, American Institute of Physics, pp 119–132
19. Nelson RE (2003) A brief history of thermophotovoltaic development. Semicond Sci Technol 18:141–143
20. Aigrain P (1956) Thermophotovoltaic conversion of radiant energy, Lecture. Massachusetts Institute of Technology, Cambridge
21. Kolm HH (1956) Solar-battery power source. Quarterly Progress Report, p 13
22. Wedlock BD (1963) Thermal photovoltaic effect. In: Proceedings of the 3rd IEEE photovoltaic specialists conference, IEEE, pp A4.1–A4.13
23. Guazzoni GE (1972) High-temperature spectral emittance of oxides of erbium, samarium, neodymium and ytterbium. Appl Spectrosc 26:60–65
24. Werth JJ (1963) Thermo-photovoltaic converter with radiant energy reflective means. US Patent 3331707
25. Nelson RE (1992) Fibrous emissive burners Selective and Broadband. Annual Report, Gas Research Inst., GRI-92/0347
26. Fraas L, Avery J, Malfa E, Wuenning JG, Kovacik G, Astle C (2003) Thermophotovoltaics for combined heat and power using low NOx gas fired radiant tube burners. In: Proceedings of the 5th conference on thermophotovoltaic generation of electricity, Rome, 16–19 Sept 2002, American Institute of Physics, pp 61–70
27. Fraas LM, Avery JE, Huang HX, Martinelli RU (2003) Thermophotovoltaic system configurations and spectral control. Semicond Sci Technol 18:165–173
28. Carlson RS, Fraas LM (2007) Adapting TPV for use in a standard home heating furnace. In: Proceedings of the 7th world conference on thermophotovoltaic generation of electricity, Madrid, 25–27 Sept 2006, American Institute of Physics, pp 273–279

29. West EM, Connelly WR (1999) Integrated development and testing of multi-kilowatt TPV generator systems. In: Proceedings of the 4th NREL conference on thermophotovoltaic generation of electricity, Denver, Colorado, 11–14 Oct 1998, American Institute of Physics, pp 446–456
30. Morrison O, Seal M, West E, Connelly W (1999) Use of a thermophotovoltaic generator in a hybrid electric vehicle. In: Proceedings of the 4th NREL conference on thermophotovoltaic generation of electricity, Denver, Colorado, 11–14 Oct 1998, American Institute of Physics, pp 488–496
31. Bitnar B, Durisch W, Mayor J-C, Sigg H, Tschudi HR, Palfinger G, Gobrecht J (2003) Record electricity-to-gas power efficiency of a silicon solar cell based TPV system. In: Proceedings of the 5th conference on thermophotovoltaic generation of electricity, Rome, Italy, 16–19 Sept 2002, American Institute of Physics, pp 18–28
32. Fraas L, Samaras J, Avery J, Minkin L (2000) Antireflection coated refractory metal matched emitters for use with GaSb thermophotovoltaic generators. In: Proceedings of the 28th IEEE photovoltaic specialists conference, pp 1020–1023
33. Fraas LM, Samaras JE, Huang HX, Minkin LM, Avery JE, Daniels WE, Hui S (2001) TPV Generators using the radiant tube burner configuration. In: Proceedings of the 17th European photovoltaic solar energy conference, Munich, 22–26 Oct 2001, WIP, pp 2308–2311
34. Wilt D, Chubb D, Wolford D, Magari P, Crowley C (2007) Thermophotovoltaics for space power applications. In: Proceedings of the 7th world conference on thermophotovoltaic generation of electricity, Madrid, 25–27 Sept 2006, American Institute of Physics, pp 335–345

Part I
Single Components

Chapter 2
Radiators (Emitters)

2.1 Introduction

Radiators (also called emitters) may be classified according to several aspects including the following:

- Optical properties (e.g., emissivity: spectral and angular, transparency).
- Thermal properties (e.g., upper operation temperature, evaporation rate, thermal expansion, thermal shock resistance and thermal conductivity).
- Electrical properties (e.g., conductivity: metal, semiconductor and non-metal).
- Material composition.
- Physical structure (e.g., bulk, porous, filaments, film and microstructured surface).
- Availability and economics.

TPV radiators have been reviewed by Coutts [1], Gombert [2], Licciulli et al. [3], Adair et al. [4] and Nelson [5]. For the selection of a radiator material, several requirements need to be considered. The radiator material needs to be thermally stable in the selected atmosphere (e.g., air, inert gas or vacuum). A good thermal shock resistance, usually achieved by low thermal expansion materials, is important for the start-up and cooling-down process of the system. High thermal conductivities result in a uniform temperature distribution of the radiators. For short wavelength (in-band) radiation it is important that the radiator has a high emissivity in order to achieve a high radiative heat transfer rate. This high rate results in a high electrical power density of the system. In addition, an appropriate radiator design also needs to consider other parameters. The bullet list of the classification shows these parameters (e.g., suitable thermal properties).

Radiators may be broadly classified in spectrally broadband and selectively emitting, or alternatively into metal and ceramic radiators. Two basic mechanisms allow spectral alteration of the emissivity. These are the material composition and the physical structure. The emissivity of radiators is not only spectrally dependent but also angle and temperature dependent.

T. Bauer, *Thermophotovoltaics*, Green Energy and Technology,
DOI: 10.1007/978-3-642-19965-3_2, © Springer-Verlag Berlin Heidelberg 2011

The following section first reviews some theory about radiator evaporation. Then the radiator concepts have been classified in broadband ceramics (Sect. 2.3), selective ceramics (Sect. 2.4) and metals (Sect. 2.5).

2.2 Thermal Stability of the Radiator

In general there is limited TPV literature available about the mass transfer from the hot radiator to the PV cell or other components (e.g., heat shield) in a TPV cavity. It was pointed out that above 1,500 K, most materials, including rare-earth oxides, show evaporation rates of a layer thickness well above 100 μm per year [6] and this could contaminate PV cells and glass shields. Fraas calculated an unacceptable deposition rate on the cell for silicon carbide radiators at 1,052°C under vacuum conditions [7]. Metal radiators (e.g., tungsten) can have an antireflection coating to enhance the emissivity in the short wavelength range. The evaporation of this coating can shift the peak of the emissivity enhancement to other wavelengths. Hence, the stability of the coating is of interest [8]. Fraas et al. considered the long-term operation of antireflection-coated tungsten radiator as feasible, if a HfO_2-coating in a noble gas at 1,250°C is utilised [7].

The Hertz–Langmuir equation has been used in the TPV community. This equation can predict the maximum evaporation rate from the solid radiator surface into a vacuum Eq. 2.1, where W is the evaporation rate (g/cm^{2}s), p_v the vapour pressure (torr)[1], T the absolute temperature (K), M the molecular weight (g/mol) and f the sticking coefficient [8–11]. For increased gas pressures, evaporation rates are lower than for vacuum conditions [11]. For TPV applications, Eq. 2.1 shows that generally radiator materials with low vapour pressure, such as tungsten, are of interest. The vapour pressure generally increases with temperature and this usually limits the upper temperature of a TPV radiator. The sticking coefficient f varies in a wide range typically from 1E-2 to 1E-6 and must be experimentally determined for each material [9].

$$W = 5.83 \cdot 10^{-2} \cdot p_v \cdot f \cdot \sqrt{M/T} \tag{2.1}$$

Knowledge about the evaporation rate is available from other research areas. For example, this mass transfer is desired for the vapour deposition of films [12]. The incandescent lamp is a good example, where undesired mass transfer could be suppressed successfully. Here, unwanted blackening of the bulb glass wall, caused by filament evaporation, needs to be minimised. Höfler, Fraas, Andreev, Luque and others realised that there are lessons to be learned from the development of light bulbs [7, 8, 13, 14].

Lamp design depends primarily on the combination of appropriate materials and the control of high-temperature transport chemistry over thousands of hours [15].

---

[1] 1 Torr = 1 mm mercury column (mm Hg) = 133.3 Pa

In an incandescent lamp, electrical energy is transformed into thermal radiation by heating a coiled tungsten wire. The temperature range varies from 1,700°C in small vacuum lamps, through 2,400°C in inert gas-filled lamps, up to 3,000°C in halogen incandescent lamps. Halogen lamps utilise a regenerative halogen (mostly iodine) cycle to return evaporated tungsten from the bulb-wall back to the filament. The pressure of the fill gas increases as the temperature of the lamp increases in order to diminish the evaporation of tungsten [15]. On the other hand, higher (noble) gas pressures reduce the efficiency due to parasitic heat losses from the hot to the cold side [8, 16]. Hence, there is a trade-off between the parasitic loss and the radiator evaporation rate. Another area of concern is the impurity level and the related transport chemistry in the incandescent lamp. The source of small quantities of impurities may be due to residual gas remaining after pumping or contaminated lamp components (e.g., water in glass). For example, residual water vapour causes wall-blackening due to the water vapour cycle [15]. Getter materials are typically utilised in order to control the chemistry of the gaseous impurities. The getters ensure that the proper gas composition is obtained and maintained during lamp life. Typically, the vapour pressure of the filaments material (e.g., tungsten) of incandescent lamps is below 0.5E-2 Pa at the temperature of operation in order to avoid wall-blackening problems caused by the physical transport mechanism [15].

Table 2.1 lists some materials with their melting points and the temperatures where the vapour pressure is 1.33E-2 Pa. This value is already higher compared to the value of 0.5E-2 Pa for a secure lamp design. The data given in Table 2.1 are sorted by the vapour pressure temperature values. The table can give some guidance for the material selection and the maximum long-term operation temperature of the radiator. The most critical materials are shown in the top half. For comparison, the melting and boiling point of the materials is also included. Most materials show a higher melting temperature compared to the critical vapour pressure temperature. Gold is a rare example where the melting temperature can be the limiting factor. In order to calculate evaporation rates, the molar mass and density values are also tabulated.

Some authors argue that most TPV systems contain a glass shield and that a contaminated shield could be replaced at fairly low cost, just like an ordinary light bulb [7, 14].

2.3 Broadband Ceramic Radiators

Ceramics tend to have fairly constant and intermediate emissivity for short wavelengths, followed by a sharp increase in wavelengths around 4–10 μm, where lattice and molecular vibration are resonant with long wavelength photons [3, 19]. The temperature dependence of the emissivity of ceramics is usually rather weak [19]. High-temperature resistant ceramics can be classified into oxide (Sect. 2.3.1) and non-oxide (Sect. 2.3.2) based.

Table 2.1 Vapour pressure and melting temperature of selected high-temperature materials sorted by the temperature for a vapour pressure of 1.33 E-2 Pa. The table was composed by the author from different literature sources [8, 9, 17, 18]

Material	Chemical symbol	Temperature (°C) for p_v = 1.33E-2 Pa	Melting point (°C)	Boiling point (°C)	Molar weight (g/mol)	Room temp. density (g/cm^3)
Silicon nitride	Si_3N_4	~800	1,900	N/A	140.284	3.2
Silicon carbide	SiC	~1,000	2,830	N/A	40.097	3.2
Silicon (IV) oxide	SiO_2	~1,025	~1,700	2,950	60.085	2.2–2.6
Gold	Au	1,132	1,064	2,856	196.967	19.3
Magnesium oxide	MgO	~1,300	2,825	3,600	40.304	3.6
Titanium (IV) oxide	TiO_2	~1,300	~1,800	~3,000	79.866	4.2
Silicon	Si	1,337	1,414	3,265	28.086	2.3
Ytterbium oxide	Yb_2O_3	~1,500	2,355	4,070	394.08	9.2
Aluminium oxide	Al_2O_3	1,550	2,054	2,977	101.961	4.0
Erbium oxide	Er_2O_3	~1,600	2,344	3,920	382.516	8.6
Holmium oxide	Ho_2O_3	n.a.	2,330	3,900	377.859	8.4
Platinum	Pt	~1,747	1,768	3,825	195.084	21.5
Yttrium oxide	Y_2O_3	~2,000	2,439	N/A	225.810	5.0
Molybdenum	Mo	~2,117	2,623	4,639	95.94	10.2
Graphite	C	2,137	#	3,825	12.011	2.2
Zirconium oxide	ZrO_2	~2,200	2,710	4,300	123.223	5.7
Hafnium oxide	HfO_2	~2,500	2,800	~5,400	210.49	9.7
Zirconium carbide	ZrC	~2,500	3,532	N/A	103.235	6.7
Tantalum	Ta	2,590	3,017	5,458	180.95	16.4
Hafnium carbide	HfC	~2,600	~3,000	N/A	190.50	12.2
Tungsten	W	2,757	3,422	5,555	183.84	19.3

At atmospheric pressure graphite vaporises without melting

2.3.1 Oxide Based Ceramics

Typical high-temperature *oxide based* ceramics are stable in oxidising atmospheres [20] and oxides include alumina (Al_2O_3), stabilised zirconia (ZrO_2), magnesia (MgO), silica (SiO_2), beryllia (BeO), hafnia (HfO_2), thoria (ThO_2) and yttria (Y_2O_3). High-temperature oxide based ceramics may consists of one or more of these oxides, such as mullite ($3Al_2O_3 \cdot 2SiO_2$), cordierite ($2MgO \cdot 2Al_2O_3 \cdot 5SiO_2$) or steatite ($MgO \cdot SiO_2$) [21]. The most commonly used high-temperature oxide based ceramic is Al_2O_3. Al_2O_3 is stable in oxidising atmosphere up to 1,900°C, which is close to its fusion temperature of ~2,050°C. Similarly ZrO_2 ceramics can withstand high-temperatures in oxidizing atmosphere and its fusion temperature is ~2,600°C [20, 21]. ZrO_2 is also examined as resistive heating material [22]. A theoretical study has considered Al_2O_3 and ZrO_2 as a TPV broadband radiator. This studied concluded that the major difficulties of these materials are their poor thermal shock resistance and the low emissivity [23].

2.3.2 Non-Oxide Based Ceramics

Common non-oxide based high-temperature ceramics are the groups of carbides (e.g., C, SiC, B_4C, WC and HfC) and nitrides (e.g., Si_3N_4, BN and AlN). Less common non-oxide based ceramics include borides (e.g., TiB_2), sulphides (e.g., CeS) and silicides (e.g., $TaSi_2$, $MoSi_2$) [8, 21–24].

Silicon nitride (Si_3N_4) can be used up to about 1,200–1,500°C in an oxidising atmosphere and has been proposed as broadband radiator for TPV [23, 24], but no TPV system development using Si_3N_4 could be identified.

Graphite (C) has a high-thermal conductivity and shows good thermal shock resistance [20]. It has a high emissivity over a wide spectral range and can be machined to the desired shape [12]. However, it can only be utilised in an oxidising atmospheres up to about 400°C [24]. Hence, graphite is of interest for non-oxidising atmospheres, such as for systems in space [7]. Graphite heat resistive elements operate in nitrogen, hydrogen and rare gases at temperatures up to 3,000°C. Electrical resistance infrared emitters based on carbon filaments with a temperature of around 1,200°C in quartz tubes are commercially available [25]. At the highest operation temperatures graphite does not melt but it sublimes [22]. The surface of graphite can be converted into SiC, which allows operation at higher temperatures in oxidising atmospheres [20].

Silicon carbide (SiC) can operate in oxidising atmospheres up to around 1,650°C due to a protective oxidation layer, which consists of silicon dioxide (passive oxidation) [20, 22, 24]. In inert gas the maximum operation temperature is lower. For example, for nitrogen a maximum temperature of 1,350°C is recommended [22]. The maximum operation temperature in vacuum is even lower. Maximum operation temperatures of 1,100°C (1E-3 Torr) and 900°C (1E-5 Torr) are stated (see also vapour pressure in Table 2.1) [22]. The oxidation resistance of SiC degrades in atmospheres containing water vapour [23]. In rare cases, silicon monoxide gas forms in atmospheres with a low partial pressure of oxygen (active oxidation) and consumes the SiC [20]. The emissivity of SiC is generally high and, in the wavelength range of interest (1–3 μm), the values range from 0.7 to 0.94 depending on author and measurement methodology [26]. The thermal conductivity of SiC is generally high, but smaller than that of graphite [20]. High-temperature application examples of SiC are heat exchangers, gas turbine components and heating elements (electric or gas heating) [20]. For example, electric SiC-heating elements were made in the US prior to the 1940s [22]. Manufacturing methods of SiC include ceramic bonded, reaction bonded, silicon nitride bonded, carbon bonded, recrystallised, sintered, hot pressed and chemical vapour deposition (CVD) [24]. SiC has been applied almost exclusively as a broadband radiator for TPV systems. The evaporation rate of SiC is high. Fraas calculated a mass transfer rate from the radiator at 1,150°C to the PV cell of 1 μm per 300 h in vacuum [7]. Guazzoni also presented some work on coated SiC radiators, which could suppress some mid-infrared radiation [27].

The discussed *silicides* are not ceramics but ceramic-metals, or cermets (e.g., $TaSi_2$, $MoSi_2$). In particular molybdenum disilicide $MoSi_2$ has been commercially used as electrical resistive heating element. $MoSi_2$ has a melting temperature of 2,050°C [21]. It is oxidation resistant due to the formation of a protective SiO_2 layer above 1,000°C in air. High maximum working temperature in oxidising atmospheres between 1,650 and 1,850°C are reported [28, 29]. Glass-bonded type $MoSi_2$ becomes plastically deformable above 1,500°C [24]. The operation temperature in vacuum is limited due to the evaporation of the protective SiO_2-layer. Brittleness of $MoSi_2$ can be also a disadvantage. For TPV emitter development, a $TaSi_2$-coating on a durable substrate such as SiC has been discussed [3].

2.4 Selective Radiators Based on Transition Metal Oxides

In the periodic table, transition metals can be further categorised into inner-transition metals (or rare-earth elements), which are made up of the lanthanide and actinide series. For selective TPV radiators some elements of the lanthanide series (*f-transition metals*) and some of the element numbers 21 to 30 (*d-transition metals*) are of interest. These elements with their incomplete filled 3d and 4f shells are marked grey in Table 2.2.

2.4.1 f-Transition Metal Oxides

The *f-transition metals* have partly filled deep lying 4f shells shielded by filled outer orbitals (Table 2.2). Therefore, electrons in the 4f shells interact little with neighbour ions and emit radiation similar to gases (line emission). Line emission is caused by discrete energy levels, which can be described by quantum mechanics. Work on the spectra and discrete energy levels of rare-earth ions in crystals have been summarised by Dieke [30, 31]. The chemical properties of the f-transition metals are similar, generally having three valence electrons ($6s^2$, $5d^1$) [3]. Guazzoni first proposed the use of high-temperature spectrally selective infrared emittance of monolithic ceramics of the oxides of Er, Sm, Nd and Yb for TPV conversion [5, 32]. Since then most work focused on Ho, Er and Yb [1]. Table 2.3 sums up the emission peak wavelengths.

2.4.2 d-Transition Metal Oxides

The *d-transition metals* have a similar unique electron configuration compared to *f-transition metals.* Their partly filled 3d orbitals are shielded by the 4s orbitals (Table 2.2). In general, the f-transition elements emit in narrower lines and are less

Table 2.2 Ground state of shell fillings, *f*- and *d-transition metals* are marked grey. For the first ten elements, the arrows show the spin directions

A. N.	Element	Shells K	L		M			N				O		P	
		1s	2s	2p	3s	3p	3d	4s	4p	4d	4f	5s	5p	5d	6s
1	H Hydrogen	↑													
2	He Helium	↑↓													
3	Li Lithium	↑↓	↑												
4	Be Beryllium	↑↓	↑↓												
5	B Boron	↑↓	↑↓	↑											
6	C Carbon	↑↓	↑↓	↑↑											
7	N Nitrogen	↑↓	↑↓	↑↑↑											
8	O Oxygen	↑↓	↑↓	↑↑↑↓											
9	F Fluorine	↑↓	↑↓	↑↑↑↓↓											
10	Ne Neon	↑↓	↑↓	↑↑↑↓↓↓											
11	Na Sodium	││	││	││││││	│										
12	Mg Magnesium	││	││	││││││	││										
13	Al Aluminium	││	││	││││││	││	│									
14	Si Silicon	││	││	││││││	││	││									
15	P Phosphorus	││	││	││││││	││	│││									
16	S Sulphur	││	││	││││││	││	││││									
17	Cl Chlorine	││	││	││││││	││	│││││									
18	Ar Argon	││	││	││││││	││	││││││									
19	K Potassium	││	││	││││││	││	││││││		│							
20	Ca Calcium	││	││	││││││	││	││││││		││							
21	Sc Scandium	││	││	││││││	││	││││││	│	││							
22	Ti Titanium	││	││	││││││	││	││││││	││	││							
23	V Vanadium	││	││	││││││	││	││││││	│││	││							
24	Cr Chromium	││	││	││││││	││	││││││	│││││	│							
25	Mn Manganese	││	││	││││││	││	││││││	│││││	││		d-transition metals					
26	Fe Iron	││	││	││││││	││	││││││	││││││	││							
27	Co Cobalt	││	││	││││││	││	││││││	│││││││	││							
28	Ni Nickel	││	││	││││││	││	││││││	││││││││	││							
29	Cu Copper	││	││	││││││	││	││││││	││││││││││	│							
30	Zn Zinc	││	││	││││││	││	││││││	││││││││││	││							
31	Ga Gallium	││	││	││││││	││	││││││	││││││││││	││	│						
32	Ge Germanium	││	││	││││││	││	││││││	││││││││││	││	││						
33	As Arsenic	││	││	││││││	││	││││││	││││││││││	││	│││						
34	Se Selenium	││	││	││││││	││	││││││	││││││││││	││	││││						
35	Br Bromine	││	││	││││││	││	││││││	││││││││││	││	│││││						
36	Kr Krypton	││	││	││││││	││	││││││	││││││││││	││	││││││						
37	Rb Rubidium	││	││	││││││	││	││││││	││││││││││	││	││││││			│			
38	Sr Strontium	││	││	││││││	││	││││││	││││││││││	││	││││││			││			
39	Y Yttrium	││	││	││││││	││	││││││	││││││││││	││	││││││	│		││			
40	Zr Zirconium	││	││	││││││	││	││││││	││││││││││	││	││││││	││		││			
41	Nb Niobium	││	││	││││││	││	││││││	││││││││││	││	││││││	││││		│			
42	Mo Molybdenum	││	││	││││││	││	││││││	││││││││││	││	││││││	│││││		│			
43	Tc Technetium	││	││	││││││	││	││││││	││││││││││	││	││││││	│││││		││			
44	Ru Ruthenium	││	││	││││││	││	││││││	││││││││││	││	││││││	│││││││		│			
45	Rh Rhodium	││	││	││││││	││	││││││	││││││││││	││	││││││	││││││││		│			
46	Pd Palladium	││	││	││││││	││	││││││	││││││││││	││	││││││	││││││││││					
47	Ag Silver	││	││	││││││	││	││││││	││││││││││	││	││││││	││││││││││		│			
48	Cd Cadmium	││	││	││││││	││	││││││	││││││││││	││	││││││	││││││││││		││			
49	In Indium	││	││	││││││	││	││││││	││││││││││	││	││││││	││││││││││		││	│		
50	Sn Tin	││	││	││││││	││	││││││	││││││││││	││	││││││	││││││││││		││	││		
51	Sb Antimony	││	││	││││││	││	││││││	││││││││││	││	││││││	││││││││││		││	│││		
52	Te Tellurium	││	││	││││││	││	││││││	││││││││││	││	││││││	││││││││││		││	││││		
53	I Iodine	││	││	││││││	││	││││││	││││││││││	││	││││││	││││││││││		││	│││││		
54	Xe Xenon	││	││	││││││	││	││││││	││││││││││	││	││││││	││││││││││		││	││││││		
55	Cs Caesium	││	││	││││││	││	││││││	││││││││││	││	││││││	││││││││││		││	││││││		│
56	Ba Barium	││	││	││││││	││	││││││	││││││││││	││	││││││	││││││││││		││	││││││		││
57	La Lanthanum	││	││	││││││	││	││││││	││││││││││	││	││││││	││││││││││	f-tr. metals	││	││││││	│	││
58	Ce Cerium	││	││	││││││	││	││││││	││││││││││	││	││││││	││││││││││	│	││	││││││	│	││
59	Pr Praseodymium	││	││	││││││	││	││││││	││││││││││	││	││││││	││││││││││	│││	││	││││││		││
60	Nd Neodymium	││	││	││││││	││	││││││	││││││││││	││	││││││	││││││││││	││││	││	││││││		││
61	Pm Promethium	││	││	││││││	││	││││││	││││││││││	││	││││││	││││││││││	│││││	││	││││││		││
62	Sm Samarium	││	││	││││││	││	││││││	││││││││││	││	││││││	││││││││││	││││││	││	││││││		││
63	Eu Europium	││	││	││││││	││	││││││	││││││││││	││	││││││	││││││││││	│││││││	││	││││││		││
64	Gd Gadolinium	││	││	││││││	││	││││││	││││││││││	││	││││││	││││││││││	│││││││	││	││││││	│	││
65	Tb Terbium	││	││	││││││	││	││││││	││││││││││	││	││││││	││││││││││	││││││││	││	││││││	│	││
66	Dy Dysprosium	││	││	││││││	││	││││││	││││││││││	││	││││││	││││││││││	││││││││││	││	││││││		││
67	Ho Holmium	││	││	││││││	││	││││││	││││││││││	││	││││││	││││││││││	│││││││││││	││	││││││		││
68	Er Erbium	││	││	││││││	││	││││││	││││││││││	││	││││││	││││││││││	││││││││││││	││	││││││		││
69	Tm Thulium	││	││	││││││	││	││││││	││││││││││	││	││││││	││││││││││	│││││││││││││	││	││││││		││
70	Yb Ytterbium	││	││	││││││	││	││││││	││││││││││	││	││││││	││││││││││	││││││││││││││	││	││││││		││
71	Lu Lutetium	││	││	││││││	││	││││││	││││││││││	││	││││││	││││││││││	││││││││││││││	││	││││││	│	││

Table 2.3 Emission peaks of *f-transition metals*

Rare earth doping species	Peak wavelength (μm)	References
Nd	2.5	[30, 32]
Sm	High emissivity from 1.8–5.0	[32]
Ho	(1.2) 2.0–2.1	[2, 3, 30]
Er	1.55	[2, 3, 5, 30, 32, 33]
Tm	1.8	[3]
Yb	0.98	[3, 5, 30, 32, 33]

affected by the host material compared to d-transition elements [34]. For TPV, cobalt (Co) and nickel (Ni) have been used with magnesia (MgO) as a host material. Suitable emission peaks at 1.13, 1.27, 1.49 μm (Co in MgO) and 1.12, 1.26, 1.41 μm (Ni in MgO) were observed [34].

2.4.3 Optically Thick Radiators

Optical thickness is defined as the product of the absorption coefficient α along a path length S. Three cases of optical thickness $\alpha(\lambda) \cdot S$ can be distinguished [35]:

- Opaque or optically thick: $\alpha(\lambda) \cdot S >> 1$.
- Transparent or optically thin: $\alpha(\lambda) \cdot S << 1$.
- Semitransparent: for all other cases of $\alpha(\lambda) \cdot S$.

Guazzoni reported about the selective emission of monolithic *f-transition metals* ceramic oxides [32]. The ceramics were *optically thick* (no flame radiation was transmitted through the sample) and erbium and ytterbium showed improved selective emission with an enhanced emissivity of around 0.6 at the peak wavelength. However, in the total wavelength range from 0.5 to 5 μm the emissivity was also generally high with values from 0.2 to 0.3 [32]. Emissivities for longer wavelengths have not been reported, but it is known that, for high-temperature resistant glasses and ceramics, the absorption coefficient typically increases considerably for wavelengths longer than about 5 μm due to lattice vibrations [3, 34, 36]. Consequently these materials also emit long wavelength radiation, which is generally undesirable for TPV operation [3]. Another difficulty with monolithic f-transition metal ceramic oxides is the poor thermal shock resistance.

Nakagawa et al. reported about an $Al_2O_3/Er_3Al_5O_{12}$ eutectic ceramic. The material was initially developed as a structural material for aircraft engines and gas turbines. For the melt growth of the material, a Bridgman-type unidirectional solidification apparatus is used. Good high-temperature performance with a thermal stability up to 1,700°C in air atmosphere has been reported. The selectivity of the emission around 1.55 μm is lower compared to the single crystal $Er_3Al_5O_{12}$ [36–40].

Although bulk *f-transition metal* ceramic oxides show some spectral selective emissivity, there is typically also a large proportion of undesirable long-wavelength radiation [4].

2.4.4 Optically Thin Radiators

Several attempts have been made to overcome thermal shock and the out-of-band emissivity challenges of monolithic radiators and they are summarised in the following.

The classical form of an *optically thin* ceramic oxide radiator using f-transition metals is the Welsbach mantle. For TPV, ytterbia (Yb_2O_3) and erbia (Er_2O_3) have been utilised and there are literature reports about high-temperature stable mantles (e.g., more than 1,700 K) with spectrally selective emission [41]. However, other sources report about scaling, fragility and durability difficulties for such Welsbach mantles [2, 3, 42, 43]. Hence various alternative optically thin ceramic oxide radiator structures, containing *f-transition metals*, have been investigated [3, 44]. These include ceramic foams [45], matrixes constructed of long fibres [3, 4, 46, 47], supported fibres [5, 48], ceramics consisting of short fibres [42], films on (polycrystalline) sapphire [49], doped silica glass [3, 50], doped garnets [51–53] and small rare-earth oxide particles suspended in a hot carrier gas and enclosed by a sapphire tube [46, 54]. Optically thin ceramic ribbons have also been made using magnesia doped with d-transition metals (cobalt and nickel) [34, 44, 55].

As already discussed, optically thin radiators must be operated in combustion zones or hot gas streams. Also, they require protective transparent shields and undesirable flame radiation is also emitted. Hence, structures on non-transparent substrates are of interest. Thermally stable ceramics with a broadband radiation (e.g., SiC or Al_2O_3) can be coated *with a porous coating containing f-transition metals*. For this approach the thickness of the coating is critical, where too small a thickness results in undesired transmission of substrate radiation through the coating [3]. On the other hand, coatings with a larger thickness can become optically thick and can have a temperature drop across the coating. In this case the volumetric emission of radiation at the lower temperature regions is usually undesirable. Erbium oxide and cobalt oxide [56], as well as porous erbium aluminium garnet [57] coatings in the order of ten to hundreds of micrometers have been applied to a SiC substrate. It has been found that scattering in the porous coatings can improve suppression of substrate radiation [3, 57].

Generally metals have inherently a desired spectral selectivity with an intermediate emittance for short wavelengths and a low emittance for long wavelengths. Optically smooth metal surfaces have a low total emittance and can be used as reflectors [56]. *Metal films*, such as platinum [36, 56] or molybdenum [8, 58], have been used *as a substrate for f- and d-transition metal ceramic oxide films*. In this configuration the oxide films are heated by conduction through the metal substrate where substrate radiation can be low because of the high metal reflectivity Oxide film materials included erbia with thulia and holmia as well as cobalt doped spinel [56], ytterbia (Yb_2O_3) [8, 58] and erbium, holmium and thulium aluminium garnets [36, 59]. For the latter, it was found that film thickness, material composition, scattering within the film and the temperature gradient across the film are all parameters that influence the (spectral) emittance of the radiator.

It has been pointed out that the emission bandwidth of a single f-transition element is relatively narrow, which could result in low electrical power densities [3]. Hence, the superposition of emissivities from different f-transition elements has been examined to broaden the spectral radiation output [3, 60].

From the literature discussed in this section, it can be concluded that in general limited work has been carried out on the durability and evaporation rate of selective ceramic radiators, but this aspect should be addressed for a durable system design.

2.5 Metal Radiators

Pure polished metals tend to have high emissivity for short wavelengths (typically 1 to 2 μm) and low emissivities in the mid and far infrared. The long-wavelength emissivity decreases approximately proportional to the square route of T/λ [19]. This natural spectral selectivity makes metals suitable for TPV radiators. On the other hand, the equation also shows that long-wavelength emissivity increases with temperature [2]. An advantage of metals is that they tend to have high thermal conductivities and this results in uniform radiator temperatures.

2.5.1 Material Options

Noble metals are known for their high resistance to oxidation in air, whereas most other metals are unstable in high-temperature oxidising atmospheres. *Gold* has a high reflectivity and does not oxidise up to its melting point of 1,064°C, but does not support its own mass at elevated temperatures [61, 62].

Elements of the *platinum group* (platinum, palladium, iridium, rhodium, osmium and ruthenium) are oxidation resistant, have high strengths at elevated temperatures and have high melting points [61]. For TPV, platinum has been used as a radiator [63]. Platinum has a melting temperature of 1,768°C. It is extremely stable and does not form protective oxide coatings. Platinum can be used in oxygen atmospheres up to 1,600°C. A common alloy is Pt–Rh which can have a higher thermal stability compared to pure Pt [22]. One major disadvantage of platinum is the high material cost.

Refractory metals have high melting points and a high thermal stability, but poor oxidisation resistance at elevated temperatures. Examples include molybdenum, tantalum and tungsten. The maximum operation temperature in air of these three metals is limited to about 500°C. All three materials are used as resistive elements in electrical heating in vacuum or protective atmospheres to prevent oxidation. In this application the maximum recommended operation temperature is 1,900, 2,200 and 2,500°C for *molybdenum, tantalum and tungsten* respectively. Molybdenum and tungsten can be used in almost any atmosphere that protects it from oxidisation. Tantalum is more critical and cannot be used in atmospheres that contain oxygen, nitrogen, hydrogen or carbon. These three metals show a good compatibility with high purity structural material such as Al_2O_3 (up to 1,900°C) and mullite (up to 1,700°C) [22]. In the TPV literature a test rig with a tantalum radiator is reported [64].

Tungsten has the highest melting point of any metal with a value of 3,422°C and it has a low vapour pressure (or low evaporation rate) [2, 8, 61]. Hence, TPV systems using tungsten as the radiator operate with a vacuum or inert gas atmosphere to avoid oxidation. A typical tungsten incandescent light bulb radiates about 8% in the visible wavelength range (0.380–0.76 μm). About 60% of the input power is infrared radiation (0.76–2.8 μm). Losses from radiation above 2.8 μm and the tube wall account for about 32%. Attempts to replace tungsten in incandescent lamps by more selective materials have failed [15]. Several TPV prototype systems using tungsten in an inert atmosphere have been developed.

The discussion shows that cost effective single elements of the metals cannot operate in oxidising environments at the temperature required for TPV emitters. Hence, *metal alloys* are of interest. Some high-temperature resistant alloys form protective oxide layers at their surface and these layers can prevent further oxidation. Commercial applications of these alloys include the areas of electrical resistance heating [28] and metal fibre gas burners. These alloys can operate at temperatures above 1,000°C in air. They are based on the major elements iron, chromium, aluminium and nickel in different compositions with trade names such as Inconcel, Rescal and Kanthal [28, 65]. An advantage of these alloys is that they can be formed by traditional metal-working methods. These alloys are also slightly spectrally selective and have a high-thermal conductivity resulting in a uniform temperature.

Different Ni–Cr and Fe–Ni–Cr alloys have been used traditionally as a resistance heating material. In this application the maximum operation temperatures is 1,100 to 1,200°C in air atmosphere. These alloys form the protective oxide CrO_2 at elevated temperatures in oxidising environments. The oxide layer thickness increases with time and spalling can be critical. Another group of alloys is based on the major elements Fe–Cr–Al in different compositions. Fe–Cr–Al alloys form a protective Al_2O_3-layer at elevated temperatures in oxidising environments. They are used up to 1,300°C in air as resistive heating element. The Al_2O_3–layer tends to adhere better compared to the CrO_2–layer. On the other hand, the emissivity of the CrO_2–layer is higher compared to the Al_2O_3–layer. In general it can be said that alloys with oxidised surfaces show different emissivities compared to non-oxidised surfaces. Some emissivities data of oxidised surfaces are available in the literature [66]. These high-temperature metal alloys, in the form of wires, have been used as a radiator up to 1,200°C directly in the combustion zone [39, 67–69]. In this way a gas fired metal radiant burner was realised. Doyle et al. reported about the use of these alloys as a structural material for the radiator tube in an indirect radiant burner for a small TPV system [63]. In summary it can be said that these commercially available alloys could be of interest for low-temperature radiators, but the maximum long-term operation temperature of these alloys, without contamination due to radiator evaporation, would need to be assessed.

2.5.2 Micro and Nano-Structures

The intermediate emissivity in the near infrared spectrum of tungsten can be enhanced using *antireflective (AR) coatings*. Theoretical emissivity values up to 95% in the wavelength range from 1 to 1.7 μm have been calculated using an AR coating thickness of about 140 nm [2]. AR coating materials are considered to have included Al_2O_3, ThO_2, ZrO_2 and HfO_2 [6, 8, 63, 70, 71]. AR coatings also allow some control of the spectral and angular emission characteristics [72]. The durability of a HfO_2-coating at 1,250°C in noble gas could also be high but this was not yet confirmed by long-term tests [7].

Surface gratings with periods in the order of micrometers have been fabricated using tungsten [2, 73–75] and other materials [76]. Several phenomena influence the spectral emissivity of these radiators including surface plasmon resonance, vertical standing waves in deep cavities and AR sub-wavelength gratings [2]. The optimum dimension of the surface gratings for high in-band and low out-of-band radiation can be found by modelling these structures (e.g., rigorous coupled-wave analyses). The durability of tungsten surface gratings is limited by surface diffusion and the gratings become almost planar after a few hours at temperatures above 1,300°C [2]. It was found that a 25 nm surface coating with HfO_2 could reduce surface diffusion considerably [2, 74]. It can be noted that tungsten lamp filaments need to retain their fibrous grain microstructure to accommodate the stresses and strains at operating temperatures of 2,200–3,100°C. This stability is achieved through additions of thoria or a combination of compounds based on aluminium, potassium and silicon [12]. The latter are also known as AKS dopants referring to the German name of the elements ("Kalium" for Potassium). Hence, one might expect that the thermal stability issues of TPV tungsten radiators at lower temperatures can be overcome using knowledge from the light bulb development.

At a fundamental research level, *photonic bandgap crystals* have been of interest for TPV conversion. In photonic bandgap crystals the lattice structure is on a scale with optical wavelengths and is based on alternating materials with high and low dielectric constants. The lattice periodicity is spaced such that bandgap wavelengths destructively interfere, thereby preventing their passage through the crystal. Constructive interference enhances wavelengths outside the bandgap and allows them to propagate through the crystal with little attenuation. This phenomenon is well known for example, for 1D multilayer films. In 1991, the first 3D photonic bandgap crystal structure was found, where photons can be blocked at the crystal bandgap regardless of the direction the photon travelled or their coherence, polarisation or angle of incidence [77]. For TPV, 3D photonic crystal radiators from tungsten have been suggested. The crystal bandgap prevents mid and far infrared radiation and the near infrared emissivities can be high [2, 78]. Photonic structures of alternating layers of ultra-thin metallic films in-between dielectric layers have been also examined as radiators [79].

It can be summarised that radiators based on micro- and nano-structures can be designed with a suitable spectrally selective performance. Critical aspects include the thermal stability and the scaling-up of the processes.

2.6 Other Novel Radiator Materials and Concepts

Other radiator materials than those summarised in the three previous subsections (broadband ceramics, selective radiators based on transition metals and metal radiators) have also been proposed. These concepts are still in a more fundamental research and development stage.

The semiconductor material *silicon* has been considered as a TPV radiator material in vacuum [8, 80]. Chubb examined theoretically and experimentally a thin, in the order of 1 μm thick, silicon film on a sapphire substrate with a reflective platinum backing film. Experiment and model approximately agreed at a temperature of 560°C and an enhanced emissivity below about 1.3 μm could be demonstrated. Considering the high vapour pressure of silicon (Table 2.1) and the vacuum operation, the maximum operation temperature would need to be assessed.

Cockeram et al. prepared, by a variety of methods, several coatings applied to molybdenum, niobium and a nickel-base alloy [9, 81]. The coatings were characterised in terms of the emissivity and the thermal stability.

Good et al. proposed to use *thermal energy storage* in between solar receiver and the radiator in order to operate the radiator also without solar radiation [82, 83]. *Latent* heat storages utilise the phase change energy, so that approximately isothermal systems can be designed. Silicon has been suggested as a phase change material (PCM) with a melting temperature of 1,414°C [83]. Fluoride salts are usually considered as a PCM in the high-temperature range in the area of latent heat thermal energy storage. An estimation of the molar latent heat can be made by Eq. 2.2, where R is the molar gas constant and T_m the melting temperature of the PCM in Kelvin. The dimensionless factor F ranges from 1 to 5 for metals, semiconductors, eutectic and inorganic compounds [84, 85]. As it can be seen from the equation, higher melting temperatures result in high melting enthalpies. Also, substances with a small molar mass are beneficial because they give higher values in terms of the melting enthalpy in the relevant unit J/g.

$$H_{m,mol} = F \cdot R \cdot T_m \tag{2.2}$$

Ashcroft and DePoy proposed *high-temperature heat pipes* (e.g., filled with lithium or sodium) to transport energy from the heat source to the radiator. This arrangement should allow a TPV system design with a high volumetric power density (W/cm^3) [86].

2.7 Summary

The various radiator materials and concepts discussed in the previous sections can be seen to be based on three technological areas.

First, *burner designs for radiant heating* have been utilised or adapted. For example metal wires arranged in the combustion zone (direct radiant burner) and recuperative single ended tube burners made of silicon carbide (indirect radiant burner) have been utilised. Section 8.4.1 discusses further details of direct and indirect radiant burners applied for heating.

Second, gas fired *radiant burner for lighting* based on the Welsbach ytterbia mantle has been adapted. The Welsbach mantel dates back to the 1890s and is also discussed in Sect. 8.4.1.

Third, as discussed in this chapter (Sect. 2.2), the tungsten radiator development profited from the *electric light bulb designs*. In general, little attention has been paid to the aspect of TPV radiator evaporation and the resulting contamination of heat shield and PV cells. It can be pointed out that such fundamental problems have been successfully handled in the tungsten light bulb operation. In addition, TPV radiators operate at much lower temperatures compared to typical light bulbs. Hence, evaporation problems are less pronounced. Although the tungsten design has some complexity due to the inert gas operation it seems a development path that has no major hindrances.

References

1. Coutts TJ, Guazzoni G, Luther J (2003) An overview of the 5th Conference on thermophotovoltaic generation of electricity. Semicond Sci Technol 18:144–150
2. Gombert A (2003) An overview of TPV emitter technologies. Proceeding of the 5th Conference on thermophotovoltaic generation of electricity, Rome, Italy, 16–19. Sep. 2002, Institute of Physics, pp 123–131
3. Licciulli A, Diso D, Torsello G, Tundo S, Maffezzoli A, Lomascolo M, Mazzer M (2003) The challenge of high-performance selective emitters for thermophotovoltaic applications. Semicond Sci Technol 18:174–183
4. Adair PL, Rose MF (1995) Composite emitters for TPV systems. Proceedings of the 1st NREL conference on thermophotovoltaic generation of electricity, Copper Mountain, Colorado, 24–28 July 1994. American Institute of Physics, pp 245–262
5. Nelson RE (1995) Thermophotovoltaic emitter development. Proceedings of the 1st NREL Conference on thermophotovoltaic generation of electricity, Copper Mountain, Colorado, 24–28 July 1994. American Institute of Physics, pp 80–96

6. Coutts TJ, Guazzoni G, Luther J (2003) An overview of the fifth Conference on thermophotovoltaic generation of electricity. Semicond Sci Technol 18:144–150
7. Fraas LM, Avery JE, Huang HX, Martinelli RU (2003) Thermophotovoltaic system configurations and spectral control. Semicond Sci Technol 18:165–173
8. Höfler H (1984) Thermophotovoltaische Konversion der Sonnenenergie (in German), Doctoral thesis. Universität Karlsruhe (TH)
9. Cockeram BV, Hollenbeck JL (1999) The spectral emittance and stability of coatings and textured surfaces for thermophotovoltaic (TPV) radiator applications, Report, US Department of Energy, DE-AC11-98PN38206
10. Roth A (1990) Vacuum technology, 3rd edn. Elsevier Science, Amsterdam
11. Darling R (2003) Micro fabrication—film deposition. Lecture notes EE-527. University of Washington College of Engineering [Online] Available at: http://www.engr.washington.edu/. Accessed 28 April 2010
12. Marsden MA, Cayless AM (1984) Lamps and lighting: a manual of lamps and lighting, 3rd edn. Routledge, London
13. Vlasov AS, Khvostikov VP, Khvostikova OA, Gazaryan PY, Sorokina SV, Andreev VM (2007) TPV Systems with solar powered tungsten emitters. Proceeding of the 7th world conference on thermophotovoltaic generation of electricity, Madrid, 25–27 Sept 2006. American Institute of Physics, pp 327–334
14. Luque A (2007) Solar Thermophotovoltaics: combining solar thermal and photovoltaics. Proceedings of the 7th world conference on thermophotovoltaic generation of electricity, Madrid, 25–27 Sept 2006. American Institute of Physics, pp 3–16
15. van den Hoek WJ, Jack AG, Luijks GMJF (2005) Lamps, in "Ullmanns Encyclopedia of Industrial Chemistry". Wiley, New York
16. Klipstein DL (2006) The great internet light bulb book, Part I [Online] Available at: http://members.misty.com/don/bulb1.html. Accessed 28 April 2010
17. Lide DR (2008-2009) Physical constants of inorganic compounds. In: CRC Handbook chemistry and physics, 89 (Eds) CRC Press
18. Global vacuum product guide (2009) Section 17-Technical Information, 9 (Eds) Kurt J Lesker Company [Online] Available at: http://www.lesker.com/newweb/literature/CDC/catalog_download.cfm. Accessed 28 April 2010
19. Modest MF (1999) Section 3.3: Radiation. In: Kreith F (ed) CRC Handbook of Thermal Engineering. CRC Press, Boca Raton, pp 65–91
20. Richerson DW (1992) Modern ceramic engineering: Properties, Processing and use in design, 2nd edn. Marcel Dekker, New York
21. Kohl WH (1967) Handbook of materials and techniques for vacuum devices. Reinhold Publishing Corporation, New York
22. Guyer EC, Brownell DL (1999) Handbook of applied thermal design. Taylor & Francis, London
23. Noreen DL, Honghua D (1995) High power density thermophotovoltaic energy conversion. Proceedings of the 1st NREL Conference on Thermophotovoltaic generation of electricity. Copper Mountain, Colorado, US, 24–28 July 1994. American Institute of Physics, pp 119–132
24. Lay LA (1991) Corrosion resistance of technical ceramics. Her Majestys Stationery Office (HMSO)
25. Infrared Emitters for Industrial Processes (2006) Heraeus [Online] Available at: http://www.noblelight.net/. Accessed 28 Oct 2010
26. Pernisz UC, Saha CK (1995) Silicon carbide emitter and burner elements for a TPV converter. Proceedings of the 1st NREL conference on thermophotovoltaic generation of electricity, Copper Mountain, Colorado, 24–28 July 1994. American Institute of Physics, pp 99–105
27. Guazzoni G, McAlonan M (1997) Multifuel (liquid hydrocarbons) TPV generator. Proceedings of the 3rd NREL Conference on Thermophotovoltaic generation of electricity. Denver, Colorado, 18–21. May 1997, American Institute of Physics, pp 341–354

28. Kanthal super electric heating element handbook (1999) Kanthal AB Sweden [Online] Available at: http://www.kanthal.com/. Accessed 28 Oct 2010
29. Watson R (1999) Electric resistance heating-element materials. In: Guyer EC, Brownell DL (eds) Handbook of applied thermal design, Part 8, Chap 1. Taylor & Francis, London
30. Chubb DL (1990) Reappraisal of solid selective emitters. Proceedings of the 21st IEEE photovoltaic specialists conference, IEEE, pp 1326–1333
31. Dieke GH (1968) Spectra and energy levels of rare earth ions in crystals. Wiley, Washington
32. Guazzoni GE (1972) High-temperature spectral emittance of oxides of erbium samarium, neodymium and ytterbium. Appl Spectrosc 26:60–65
33. Touloukian YS, DeWitt DP (1972) Thermophysical properties of matter, Vol. 8, Thermal radiative properties: nonmetallic solids. Plenum Press, New York
34. Ferguson LG, Dogan F (2002) Spectral analysis of transition metal-doped MgO matched emitters for thermophotovoltaic energy conversion. J Mater Sci 37(7):1301–1308
35. Siegel R, Howell J (2001) Thermal radiation heat transfer, 4th edn. Taylor & Francis, London
36. Chubb DL, Pal A-MT, Patton MO, Jenkins PP (1999) Rare earth doped high-temperature ceramic selective emitters. J Eur Ceramic Society 19:2551–2562
37. Yugami H, Sai H, Nakamura K, Nakagawa N, Ohtsubo H (2000) Solar thermophotovoltaic using $Al_2O_3/Er_3Al_5O_{12}$ eutectic composite selective emitter. Proceedings of the 28th IEEE photovoltaic specialists conference, IEEE, pp 1214–1217
38. Adachi Y, Yugami H Shibata K, Nakagawa N (2004) Compact TPV generation system using Al2O3/Er3Al5O12 eutectic ceramics selective emitters. Proceedings of the 6th international conference on thermophotovoltaic generation of electricity, Freiburg, Germany, 14–16 June 2004. American Institute of Physics, pp 198–205
39. Mattarolo G (2007) Development and modelling of a thermophotovoltaic system. Doctoral thesis, University of Kassel
40. Nakagawa N, Ohtsubo H, Waku Y, Yugami H (2005) Thermal emission properties of Al2O3/ Er3Al5O12 eutectic ceramics. J Eur Ceramic Society 25(8):1285–1291
41. Bitnar B, Durisch W, Mayor J-C, Sigg H, Tschudi HR (2002) Characterisation of rare earth selective emitters for thermophotovoltaic applications. Sol Energy Mater Sol Cells 73:221–234
42. Diso D, Licciulli A, Bianco A, Lomascolo M, Leo G, Mazzer M, Tundo S, Torsello G, Maffezzoli A (2003) Erbium containing ceramic emitters for thermophotovoltaic energy conversion. Mater Sci Eng 98(2):144–149
43. Palfinger G (2006) Low dimensional Si/SiGe structures deposited by UHV-CVD for thermophotovoltaics. Doctoral thesis, Paul Scherrer Institute
44. Ferguson LG, Dogan F (2001) Spectrally selective, matched emitters for thermophotovoltaic energy conversion processed by tape casting. J Mater Sci 36(1):137–146
45. Schubnell M, Gabler H, Broman L (1997) Overview of European activities in thermophotovoltaics. Proceedings of the 3rd NREL Conference on thermophotovoltaic generation of electricity. Denver, Colorado, 18–21 May 1997. American Institute of Physics, pp 3–22
46. Ortabasi U, Lund KO, Seshadri K (1996) A fluidized bed selective emitter system driven by a non-premixed burner. Proceedings of the 2nd NREL conference on thermophotovoltaic generation of electricity, Colorado, Springs, 16–20 July 1995. American Institute of Physics, pp 469–487
47. Adair PL, Zheng-Chen, Rose F (1997) TPV power generation prototype using composite selective emitters. Proceedings of the 3rd NREL Conference on thermophotovoltaic generation of electricity, Denver, Colorado, 18–21 May 1997. American Institute of Physics, pp 277–291
48. Nelson RE (1997) Temperature measurement of high performance radiant emitters. Proceedings of the 3rd NREL conference on thermophotovoltaic generation of electricity, Denver, Colorado, 18–21 May 1997. American Institute of Physics, pp 189–202
49. Pierce DE, Guazzoni G (1999) High-temperature optical properties of thermophotovoltaic emitter components. Proceedings of the 4th NREL conference on thermophotovoltaic generation of electricity, Denver, Colorado, 11–14 Oct 1998. American Institute of Physics, pp 177–190

50. Licciulli A, Maffezzoli A, Diso D, Tundo S, Rella M, Torsello G, Mazzer M (2003) Sol-gel preparation of selective emitters for thermophotovoltaic conversion. J Sol-Gel Sci Technol 26(1):1119–1123
51. Panitz J-C, Schubnell M, Durisch W, Geiger F (1997) Influence of ytterbium concentration on the emissive properties of Yb:YAG and Yb:Y_2O_3. Proceedings of the 3rd NREL conference on thermophotovoltaic generation of electricity, Denver, Colorado, 18–21 May 1997. American Institute of Physics, pp 265–276
52. Panitz J-C (1999) Characterization of ytterbium-yttrium mixed oxides using Raman spectroscopy and x-ray powder diffraction. J Raman Spectrosc 30(11):1035–1042
53. Goldstein MK, DeShazer LG, Kushch AS, Skinner SM (1997) Superemissive light pipe for TPV applications. Proceedings of the 3rd NREL conference on thermophotovoltaic generation of electricity, Denver, Colorado, 18–21 May 1997. American Institute of Physics, pp 315–326
54. Chubb DL, Lowe RA (1996) A small particle selective emitter for thermophotovoltaic energy conversion. Proceedings of the 2nd NREL conference on thermophotovoltaic generation of electricity, Colorado Springs, 16–20. July 1995. American Institute of Physics, pp 263–277
55. Ferguson L, Fraas L (1997) Matched infrared emitters for use with GaSb TPV cells. Proceedings of the 3rd NREL conference on thermophotovoltaic generation of electricity, Denver, Colorado, 18–21 May 1997. American Institute of Physics, pp 169–179
56. Crowley CJ, Elkouh NA, Magari PJ (1999) Thermal spray approach for TPV emitters. Proceedings of the 4th NREL conference on thermophotovoltaic generation of electricity, Denver, Colorado, 11–14 Oct 1998. American Institute of Physics, pp 197–213
57. Diso D, Licciulli A, Bianco A, Leo G, Torsello G, Tundo S, Sinisi M, Larizza P, Mazzer M (2003) Selective emitters for high efficiency TPV conversion: Materials preparation and characterisation. Proceedings of the 5th conference on thermophotovoltaic generation of electricity, Rome, 16–19 Sept 2002. American Institute of Physics, pp 132–141
58. Höfler H, Würfel P, Ruppel W (1983) Selective emitters for thermophotovoltaic solar energy conversion. Solar Cells 10(3):257–271
59. Good BS, Chubb DL (2003) Theoretical comparison of erbium-, holmium- and thulium-doped aluminum garnet selective emitters. Proceedings of the 5th conference on thermophotovoltaic generation of electricity, Rome, Italy, 16–19 Sept 2002. American Institute of Physics, pp 142–154
60. Zheng C, Adair PL, Rose MF (1997) Multiple-dopant selective emitter. Proceedings of the 3rd NREL conference on thermophotovoltaic generation of electricity, Denver, Colorado, 18–21 May 1997. American Institute of Physics, pp 181–188
61. Wyatt L (1993) Materials properties and selection. In: Koshal D (ed) Manufacturing engineers reference book, Chap 1, 13th edn. Butterworth-Heinemann, London
62. Korb LJ (1987) Metals handbook Vol 13 corrosion, 9th edn. ASM International, New York
63. Doyle E, Shukla K, Metcalfe C (2001) Development and demonstration of a 25 W thermophotovoltaic power source for a hybrid power system, Report, NASA, TR04-2001
64. Khvostikov VP, Gazaryan PY, Khvostikova OA, Potapovich NS, Sorokina SV, Malevskaya AV, Shvarts MZ, Shmidt NM, Andreev VM (2007) GaSb Applications for solar thermophotovoltaic conversion. Proceedings of the 7th world conference on thermophotovoltaic generation of electricity, Madrid, 25–27 Sept 2006. American Institute of Physics, pp 139–148
65. (2010) Special Metals Corporation, Product Inconel [Online] Available at http://www.specialmetals.com/. Accessed 28 Oct 2010
66. Wheeler MJ (1983) Radiating properties of metals. In: Brandes EA (ed) Smithells metals reference book, Chap 17, 6th edn. Butterworth-Heinemann, London
67. Zenker M (2001) Thermophotovoltaische Konversion von Verbrennungswaerme (in German), Doctoral thesis, Albert-Ludwigs-Universität Freiburg im Breisgau
68. Volz W (2001) Entwicklung und Aufbau eines thermophotovoltaischen Energiewandlers (in German), Doctoral thesis, Universität Gesamthochschule Kassel, Institut für Solare Energieversorgungstechnik (ISET)

69. Rumyantsev VD, Khvostikov VP, Sorokina O, Vasilev AI, Andreev VM (1999) Portable TPV generator based on metallic emitter and 1.5-amp GaSb cells. Proceedings of the 4th NREL conference on thermophotovoltaic generation of electricity, Denver, Colorado, 11–14 Oct 1998. American Institute of Physics, pp 384–393
70. Fraas LM, Magendanz G, Avery JE (2001) Antireflection coated refractory metal matched emitters for use in thermophotovoltaic generators. JX-Crystals Inc., US Patent 6177628
71. Fraas LM, Samaras JE, Avery JE (2001) Antireflection coated refractory metal matched emitters for use in thermophotovoltaic generators. JX-Crystals Inc., US Patent 6271461
72. Les J, Borne T, Cross D, Gang Du, Edwards DA, Haus J, King J, Lacey A, Monk P, Please C, Hoa Tran (2000) Interference filters for thermophotovoltaic applications. Proceedings of the 15th workshop on mathematical problems in industry, University of Delaware, US, June 1999
73. Sai H, Yugami H, Kanamori Y, Hane K (2003) Spectrally selective emitters with deep rectangular cavities fabricated with fast atom beam etching. Proceedings of the 5th conference on thermophotovoltaic generation of electricity, Rome, Italy, 16–19 Sept 2002. American Institute of Physics, pp 155–163
74. Schlemmer C, Aschaber J, Boerner V, Gombert A, Hebling C, Luther J (2003) Thermal stability of microstructured selective tungsten emitters. Proceedings of the 5th Conference on thermophotovoltaic generation of electricity, Rome, Italy, 16–19 Sept 2002. American Institute of Physics, pp 164–173
75. Sai H, Kamikawa T, Yugami H (2004) Thermophotovoltaic generation with microstructured tungsten selective emitters. Proceedings of the 6th international conference on thermophotovoltaic generation of electricity, Freiburg, Germany, 14–16 June 2004. American Institute of Physics, pp 206–214
76. Pralle MU, Moelders N, McNeal MP, Puscasu I, Greenwald AC, Daly JT, Johnson EA, George T, Choi DS, El-Kady I, Biswas R (2002) Photonic crystal enhanced narrow-band Infrared emitters. Appl Phys Lett 81:4685–4687
77. McCarthy DC (2002) Photonic crystals: a growth industry, Photonics Spectra, June, pp 54–60
78. Fleming JG, Lin SY, El-Kady I, Biswas R, Ho KM (2002) All-metallic three-dimensional photonic crystals with a large infrared bandgap. Nature 417:52–55
79. Narayanaswamy A, Cybulski J, Gang Chen (2004), 1D metallo-dielectric photonic crystals as selective emitters for thermophotovoltaic applications. Proceeding of the 6th international conference on thermophotovoltaic generation of electricity, Freiburg, Germany, 14–16 June 2004. American Institute of Physics, pp 215–220
80. Chubb DL, Wolford DS, Meulenberg A, DiMatteo RS (2003) Semiconductor silicon as a selective emitter. Proceedings of the 5th conference on thermophotovoltaic generation of electricity, Rome, Italy, 16–19 Sept 2002. American Institute of Physics, pp 174–200
81. Cockeram BV, Measures DP, Mueller AJ (1999) The development and testing of emissivity enhancement coatings for themophotovoltaic (TPV) radiator applications. Thin Solid Films 355–356:17–25
82. Good BS, Chubb DL, Lowe RA (1997) Comparison of selective emitter and filter thermophotovoltaic systems. Proceeding of the 2nd NREL conference on thermophotovoltaic generation of electricity, Colorado Springs, 16–20 July 1995. American Institute of Physics, pp 16–34
83. Chubb DL, Good BS, Lowe RA (1996) Solar thermophotovoltaic (STPV) system with thermal energy storage. Proceedings of the 2nd NREL conference on thermophotovoltaic generation of electricity, Colorado, Springs, 16–20 July 1995. American Institute of Physics, pp 181–198
84. Dincer I (2002) Thermal energy storage systems as a key technology in energy conservation. Int J Energy Res 26:567–588
85. Hahne E (2005) Heat Storage Media, In: Ullmanns encyclopedia of industrial chemistry. Wiley, New York
86. Ashcroft J, DePoy D (1997) Design considerations for a thermophotovoltaic energy converter using heat pipe radiators, Report, Kapl Atomic Power Laboratory, US, KAPL-P-000236

Chapter 3
Filters

3.1 Introduction

Overviews of infrared filters are given by Good et al. [1], Chubb et al. [2, 3], Gruenbaum et al. [4], Horne et al. [5] and Köstlin [6]. Generally, TPV systems with filters should achieve a higher efficiency and power density compared to available selective radiators as modelling by Good et al. showed [1]. Two general types of ideal filters have been proposed (Fig. 3.1) [4, 7]. Both filters ideally fully reflect photons with energies below the PV cell bandgap energy; or in other words long wavelength radiation (out-of-band) should be reflected. For high efficiencies, usually out-of-band reflectivity of more than 90% would be ideal assuming currently used cell bandgaps and radiator temperatures, as pointed out by Baldasaro et al. [8]. A similar conclusion is also drawn in Sect. 6.2.3. Also, both filters show ideally no absorption of radiation at all wavelengths.

The first filter is a very narrow band pass that transmits photons with energies that are slightly above the PV cell bandgap energy; all other photons would be reflected. Such band-pass filter results in an ideal conversion of the photons in the PV cell without excess photon energy and leads to the highest conversion efficiency. On the other hand, a narrow wavelength band contains only a limited number of photons and this results in a low power density.

The second filter is an edge filter that transmits all photons with energies above the PV cell bandgap and reflects all photons below the bandgap energy. In terms of radiation, short wavelength radiation (in-band) is transmitted and long wavelength radiation (out-of-band) is reflected. The selection of these two ideal filters is a choice between either a high power density or a high efficiency. In other words, it is not possible to maximise both the power density and the efficiency.

Most practical systems aim for an edge filter rather than a band-pass filter design. It is also the opinion of the author that an edge filter is more advantageous as discussed in Sect. 6.2.3. Results of this modelling show that a band-pass filter leads to a large power density drop and the gain in the efficiency is not very high compared to the edge filter for typical configurations currently considered.

T. Bauer, *Thermophotovoltaics*, Green Energy and Technology,
DOI: 10.1007/978-3-642-19965-3_3, © Springer-Verlag Berlin Heidelberg 2011

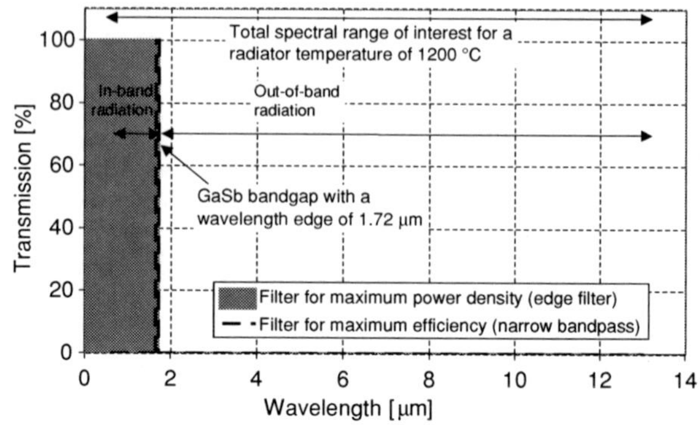

Fig. 3.1 Example of the transmission of an ideal filter for maximum power density (*grey transmission area*) and efficiency (*dashed black line*) of a TPV system using a GaSb PV cell

Table 3.1 Radiation range and peak wavelength for different blackbody temperatures, as well as the total power of a blackbody

Blackbody temperature (°C)	Total blackbody radiation (W/cm^2)	Lower 1% edge of radiation (μm)	Peak wavelength (μm) (Wien's displacement law)	Upper 1% edge of radiation (μm)
800	7.5	0.89	2.70	17.9
1,000	14.9	0.75	2.28	15.1
1,200	26.7	0.65	1.97	13.0
1,400	44.4	0.57	1.73	11.5
1,600	69.8	0.51	1.55	10.2
1,800	104.7	0.46	1.40	9.3

For an optimum filter design, different aspects need to be considered. Table 3.1 shows that the filter parameters depend on the radiator (or blackbody) temperature. The 1% edges are defined as the wavelengths where the blackbody power peak value drops to 1% of its value. The integrated radiation from the lower to the upper wavelength edge, given in Table 3.1, is 98.5% of the total blackbody power. Hence, for spectral control the lower and upper edge define the total wavelength range of interest. This total range becomes smaller for higher radiator temperatures. For example, the upper edge of radiation changes from about 18 μm at 800°C to about 9 μm at 1,800°C. Although bodies slightly visible glow by incandescence below 800°C, it can be seen that at 800°C almost all radiation is outside the visible spectral range (0.38–0.76 μm).

For TPV applications, the peak wavelength of the blackbody is typically in the long wavelength (out-of-band) spectral range. The filter performance in terms of a high reflectivity should be best close to this peak wavelength, since the radiation intensity is highest around the peak. Another issue is the strong increase of blackbody radiation directly proportional to the fourth power of the absolute temperature. As a consequence, parasitic absorption by the filter and filter cooling will play a more important role at higher radiator temperatures.

Another general aspect is the angle of incidence of the radiation. Radiators in a TPV system typically emit radiation according to a Lambertian radiation distribution and the radiation on the filter surface arrives under a wide range of incidence angles. Hence, filters need to show a high performance for a wide range of angles.

It should be pointed out that knowledge about infrared reflective filters is available from other applications [6]. Examples are the thermal insulation of light bulbs, spectrally selective solar absorbers and double-glazed building windows. For example, the efficiency of an electric light bulb can be increased significantly, if infrared radiation is reflected back to the source and reabsorbed by the filament. Spectrally selective solar absorbers have been realised with a high solar radiation absorption and high thermal (infrared) reflectance of the hot absorber (or low emittance) [9, 10]. In double-glazed windows about two-third of the heat loss occurs via thermal radiation and this loss can be minimised by coating the inner surface or surfaces with an infrared reflective filter [6].

In general, the filter requirements for an efficient TPV system (wide spectral range, edge filter characteristic with little absorption, angle characteristic) are not easy to meet. Hence, various filter technologies including frequency selective surface (FSS), transparent conducting oxide (TCO), all-dielectric and metal-dielectric filters are of interest (Sects. 3.3–3.6). Also filters that combine the advantages of different filter types are developed (Sect. 3.7 Dielectric-TCO). Another spectral control option is the back surface reflector (BSR) as an integral part of the PV cell (see Sect. 4.4).

Most TPV systems utilise a bulk dielectric (e.g., quartz glass) shield in the cavity. In general, these glass heat shields cannot fulfil the filter requirement in terms of the spectral selectivity. Nevertheless, heat shields are included in this chapter, since they can contribute to some of the spectral control in a TPV system.

3.2 Bulk Dielectrics in the Cavity (Heat Shields)

Many TPV systems use a transparent high-temperature heat shield between the radiator and the PV cells. Such heat shields perform several roles including the following:

- Minimisation of conduction and convection heat transfer between radiator and PV cell,
- Maintenance of an inert atmosphere or vacuum around metal radiators to avoid radiator oxidation (e.g., tungsten),
- Protection of PV cells from combustion products of direct radiant burners.

First, as a potential heat shield option crystalline and polycrystalline materials are discussed (Sect. 3.2.1), although almost exclusively fused silica, or quartz glass (SiO_2), has been used (Sect. 3.2.2).

3.2.1 Crystalline Materials

Sapphire (crystalline Al_2O_3) has a high melting point of 2,054°C. The transparency range is approximately from 0.2 to 5 μm with a very low absorption

Table 3.2 Crystalline infrared optical materials with a high melting point, sorted by vapour pressure and then melting point. The table was composed by the author from several literature sources [13, 15–17]

Material	Chemical Symbol	Melting point (°C)	Temperature for a vapour pressure of 1.33 E-2 Pa (°C)	Transparency range (μm)
Boron phosphide	BP	1,125 decomp.	N/A	0.5–N/A
Barium titanate	$BaTiO_3$	1,625	N/A	N/A
Yttrium aluminium oxide	$Y_3Al_5O_{12}$	1,930	N/A	0.2–5
Strontium titanate	$SrTiO_3$	2,080	N/A	0.5–5
Spinel	$MgAl_2O_4$	2,135	N/A	0.2–5
Aluminium Oxynitride	$Al_{23}O_{27}N_5$	2,170	N/A	0.2–5
Diamond	C	~3,500	N/A	0.2–3
Zinc selenide	ZnSe	1,520	660	0.5–19
Zinc sulphide (wurtzite)	ZnS	1,700	800	0.4–12.5
Silicon nitride	Si_3N_4	1,900	800	N/A
Gallium phosphide	GaP	1,457	920	0.5–N/A
Lanthanum fluoride	LaF_3	1,493	900	0.1–10
Silicon carbide	SiC	2,830	1000	0.5–4
Magnesium oxide	MgO	2,825	1,300	0.4–7
Titanium oxide	TiO_2	1,560–1,843	1,300	0.4–4
Silicon	Si	1,414	1,337	1.1–6.5
Sapphire	Al_2O_3	2,054	1,550	0.2–5
Boron nitride	BN	2,967	1,600	0.2–N/A
Aluminium nitride	AlN	3,000	1,750	N/A
Yttria	Y_2O_3	2,439	2,000	0.3–7

coefficient in the range from 1 to 3.3 μm. The long-term operation limit is below 1,500°C due to the vapour pressure (Table 3.2). It is known that alumina powder of high purity (i.e., >99.9%) can be also sintered to translucency by addition of small amounts of sinter additives (e.g., about 0.2 wt% MgO); sintering is typically performed at temperatures of 1,827°C to obtain *polycrystalline alumina* (PCA). PCA is used as arc tube material in lamps. PCA is a cost effective alternative, but has a reduced transmission due to scattering effects compared to single crystal alumina [11].

Other infrared optical materials, with a high thermal stability and a vapour pressure lower than sapphire, include boron nitrite, aluminium nitride and yttria. *Yttrium aluminium garnet* (YAG, $Y_3Al_5O_{12}$) has also a high melting point of approximately 1,930°C [12, 13]. YAG is commonly used as a host material in solid-state lasers. Goldstein et al. used YAG as a light pipe to guide the radiation by total internal reflection from a high-temperature heat source to the PV cell [12]. Chubb modelled a light pipe concept where a heat transfer enhancement proportional to the refractive index squared can be achieved (see Sect. 6.5.2). In the theoretical work as a candidate material ZnSe was used [14]. ZnSe has a very wide transmission range from 0.5 to 19 μm. However, the high vapour pressure of ZnSe could prohibit long-term operation at high temperatures (Table 3.2).

As will be discussed in Sect. 3.2.2, absorption and reemission in the heat shield of long wavelength radiation above the bandgap allows for some limited spectral control. Table 3.2 shows that almost all crystalline materials are transparent (or have a low absorption) up to high wavelengths, if compared to fused silica with a smaller transparency window. In other words, most crystalline materials would transmit long wavelength (out-of-band) radiation, whereas fused silica would absorb part of this radiation. The fused silica shield also reemits part of the radiation towards the emitter. In this way radiation is returned to the emitter. Another major disadvantage of single crystalline materials is usually their higher costs due to the growth process compared to fused silica.

3.2.2 Amorphous Materials (Glasses)

Amorphous materials (glasses) fulfil many of the required TPV heat shield characteristics. Quartz glass (fused silica or SiO_2) has been almost exclusively used [18–21]. Although, high-temperature borosilicate glasses with tradenames such as Duran® [20] and Pyrex® [19] have also been considered.

Table 3.3 lists some important properties of glasses used in electric lamps [11]. The advantages of soda lime glass include low cost and easy processability. On the other hand, the heat and thermal shock resistance of this soft glass is relatively poor. Hard glasses, such as borosilicate and aluminosilicate glasses have lower thermal expansion coefficients, higher softening temperatures and they are better suited for high-temperature usage. High silica glasses have the highest maximum operation temperatures and very good resistance against heat and thermal shock (low thermal expansion coefficient). The drawbacks of pure quartz glass are its high cost and high processing temperature. Hence, cheaper alternative high silica glasses have been developed and they could also be of interest for TPV systems. On the one hand, there is vycor glass that contains a few percent B_2O_3. On the other hand, doping of pure SiO_2 melt with a few hundred ppm of oxides like BaO, Al_2O_3 and K_2O reduces the processing temperature. For tungsten halogen lamps doped quartz glass up to a typical operation temperature of 830°C can be employed.

For very high-temperature applications pure quartz glass has typically the appropriate thermal properties. It has the highest melting point of any glass and has a good thermal shock resistance [13, 22]. At very high temperatures *devitrification* can occur [22]. Here the white opaque cristobalite (a crystalline phase of SiO_2) forms in the SiO_2 surface layer. Cristobalite has a much higher expansion coefficient than quartz glass. This acts as a stress raiser and results in spalling. Hence, the cristobalite layer proceeds inwards from the surface [11, 22, 23]. There are various factors influencing the onset temperature and the rate of devitrification, such as the SiO_2 type, operation temperature, exposure time, thermal history and surface contamination [24]. Especially impurities from alkali metal ions accelerate

Table 3.3 Properties of selected (high temperature) glasses, adapted from [11]

Glass type	Typical composition (wt%) (empty fields refer to minor constituents)								Softeing point (°C)	Typical max. temp. (°C)	Expansion coeff. 0–300°C (10–6K^{-1})
	SiO_2	Na_2O	K_2O	B_2O_3	Al_2O_3	MgO	CaO	BaO			
Soft glass											
Soda lime	72	16	1	–	2	4	5	–	700	N/A	9.4
Hard glass											
Borosilicate	78	5	–	15	2	–	–	–	800	N/A	4
Aluminosilicate	61	–	–	1	16	–	10	12	1,025	680	4.5
High silica glass											
Vycor quartz	96	–	–	4	–	–	–	–	1,020	930	0.75
Pure quartz	>99.9	–	–	–	–	–	–	–	1,140	1,130	0.55

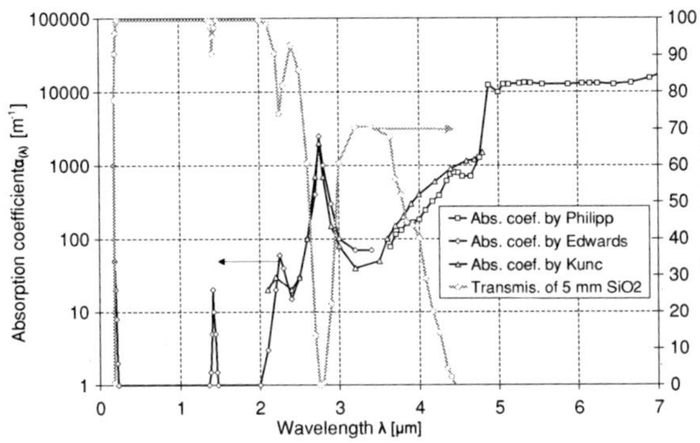

Fig. 3.2 Absorption coefficient of non-water free quartz glass (*left axis*). This data was used to calculate the transmission for a path length of 5 mm (*right axis*)

the onset and the rate of devitrification [11]. Quartz glass can be used up to 1,130°C, although devitrification (i.e., crystallization) may occur from 1,000°C upwards [11].

Pure quartz glass is available in different *grades* and these can be classified in electrically fused (or water-free or infrared-grade), flame-fused and vitreous synthetic silica [24]. The grades differ in their hydroxyl content: electrically fused (low), flame-fused (intermediate) and vitreous synthetic (high) [24]. Water in the form of hydroxyl groups in the quartz glass causes absorption peaks [24]. Absorption peaks occur at wavelengths of 1.38, 2.22 and 2.73 µm [24]. Assuming a quartz glass thickness in the range of centimetres, three wavelength intervals for the transmission can be distinguished for quartz glass. These are below 2 µm (optically-thin), from 2 to 5 µm (semitransparent) and above 5 µm (optically-thick). Figure 3.2 shows the absorption coefficient $\alpha_{(\lambda)}$ of non-infrared-grade quartz glass. The data were taken from various sources [25, 26]. The absorption coefficient $\alpha_{(\lambda)}$ for longer wavelengths was calculated from the extinction coefficient $k_{(\lambda)}$ given by Philipp using the relation $\alpha_{(\lambda)} = 4\pi \cdot k_{(\lambda)}/\lambda$ [27, 28]. The quartz glass absorption coefficient up to $\lambda = 2$ µm is very low (Fig. 3.2). The absorption coefficient values in the window from around 1.2 to 1.6 µm are lower than 1 m^{-1}. This spectral window is utilised for long-distance optical fibre transmission [28].

There have been a few TPV publications assessing how SiO_2 shields situated between radiator and PV cell influence the radiative heat transfer. Modelling by Hottel [29] pointed out that the glass shield acts as a thermal protection for the PV cells but also reduces convertible short wavelength transmission in the cavities. This modelling has not included reemission of radiation of these shields. Experimental work by Pierce et al. [18] and Fraas et al. [30] indicates that re-emission of the hot shields needs to be taken into account. It has been pointed out that hot shield reemits radiation in both direction, to the radiator and the PV cell [18, 30]. This allows suppression of some undesirable long wavelength radiation, but the quality of this suppression can be regarded as limited and does not fulfil the requirements of a TPV filter for a highly efficient system. Considering different quartz glass types, non-water-free quartz glass seems to fulfil the spectral requirements best. There is a small, usually undesirable, absorption peak at

1.38 μm and stronger desired long wavelength absorption bands around the peaks at 2.22 and 2.73 μm [24]. The long wavelength radiation will be absorbed, at least partly by the water bands, and reemitted in both directions (to the radiator and the cell). This will contribute to the spectral control in the system. In comparison, crystalline materials cannot absorb long wavelength radiation due to their usually wide window (compare Table 3.2).

External cooling can keep the shield at a lower temperature to minimise re-emission of radiation. However, this can be regarded as generally undesirable, since this absorbed energy in the shield is lost and degrades the overall efficiency of the TPV system. Hence, operation of the heat shield in radiative equilibrium with its environment (radiator, PV cells) is usually of interest.

Another suggested option is the use of *multiple shields* to improve the suppression of long wavelengths [30–33]. Fraas concluded that N shields reduce long wavelength radiation by 1/(N + 1) [30]. However for multiple shields it has been found that not only long wavelength radiation, but also desirable short wavelength radiation is reduced by reflection losses [18, 31–34]. The reduced short wavelength transmission leads to low electrical power densities and this is usually undesirable. In order to reduce the reflection losses, the utilisation of antireflection coating has been proposed [32].

Modelling by Hanamura [31–33] suggests that an *increased shield thickness* in the order of centimetres can improve long wavelength suppression. However, in their work the SiO_2 shield was not in radiative equilibrium since convective heating and cooling from flue gas and combustion-air was anticipated. Modelling and experimental work in radiative equilibrium performed by the author showed that an increased SiO_2 shield thickness suppresses long wavelength more effectively than thin shields. The modelling also showed that the shield materials utilised as filters should have a high absorption coefficient in the long wavelength range and a low thermal conductivity. This suppression can be attributed to a temperature gradient and the coupled radiation and conduction heat transfer in the glass. Section 5.4.5 discusses the theory of coupled radiative and conductive heat transfer in semitransparent media (e.g. glass) [34].

In general, it can be pointed out that heat shields made of quartz glass can minimise conduction and convection heat transfer from the radiator to the PV cell. For radiators such as tungsten quartz glass can provide the envelope for the inert gas atmosphere. Also, the quartz shield can contribute to some spectral control, but the quality of spectral control, without other spectral control measures, is not sufficient for an efficient system design. For these reasons, quartz glass shields are often an integral and important part of typical TPV systems.

3.3 Frequency Selective Surface (FSS) Filters

Frequency selective surface (FSS) filters are essentially array structures that consist of a plurality of thin conducting elements, often printed on a dielectric substrate for support [35]. Examples of element shapes are dipole, tripole,

Jerusalem cross, square loop, cross dipole and circular loop [35]. Frequently, FSSs form band-pass filters, also called resonant array or metallic mesh filters, which consist of a conducting plate with periodic apertures (inductive type). The complementary structure with periodic metal patches forms a band-rejection filter (conductive type) [5]. FSSs are well known for the sub millimetre and micrometer electromagnetic wavelength range [7]. For example, they are familiar from the door of a common microwave oven where they reflect long wavelength electromagnetic radiation (microwaves) and transmit short wavelength radiation (light) [36].

For TPV, band-pass filters are of interest. These filters reflect most radiation except photons with energies in a band above the PV cell bandgap energy. The challenge of these filters applied for the near infrared wavelength range is the fabrication of the physically small structures with dimensions smaller than the peak transmission wavelength [37]. Hence, nanofabrication techniques such as electron beam lithography, ion beam lithography or nanoimprint lithography are required [36, 38].

Horne at EDTEK Inc. has developed inductive FSSs using about 420 million cross dipoles per cm^2 [5, 39–42]. Initially each dipole has been fabricated individually by direct-electron-beam lithography, which has been replaced by a masked-ion-beam lithography process to reduce costs. The peak transmission at resonance frequency has been optimised for a wavelength of 1.45 μm for GaSb cells responding up to 1.8 μm. Disadvantages of this filter include high costs and the transmission versus wavelength performance (approximately Gaussian shape). The transmission characteristic leads to an in-band transmission of less than 100% that reduces the power density. The Gaussian shape transmission characteristic also has no sharp cut-off wavelength so that some out-of-band transmission occurs. Nevertheless, systems using this filter are among the most efficient TPV prototypes being built. This filter has been incorporated or considered for combustion, solar and nuclear TPV systems [40, 43]. The high performance is achieved due to a high reflectivity of mid and long wavelength radiation similar to a gold mirror (98–99% reflectivity) [5] and the suitable angular performance of the filter. The transmission peak wavelength (or resonant frequency) of this filter remains the same for incident radiation with large zenith angles, although the filter increasingly reflects this radiation (specular surface behaviour) [5].

Kristensen et al. fabricated and examined also other element structures than the cross dipole for TPV FSS filters [44]. They concluded that inherent parasitic absorption, which is the result of the induced currents in the FSS metallisation, is a significant obstacle to achieve the very high performance of combined dielectric–TCO filters (see Sect. 3.7).

3.4 Transparent Conducting Oxide (TCO) Filters

TCOs, or semiconductor or plasma filters, are highly doped thin semiconductor films with a large bandgap [6, 7]. Examples include indium oxide (In_2O_3), tin oxide (SnO_2), indium tin oxide (or ITO, In_2O_3–SnO_2), zinc oxide (ZnO), cadmium stannate (or CTO, Cd_2SnO_4) and cadmium indium oxide (or CIO, $CdO–In_2O_3$). For In_2O_3 and ITOs typically Sn is used as a dopant [6, 45]. For TPV conversion, arsenic, phosphorus and boron doped silicon TCOs were also examined [46].

There are different known deposition methods and they include chemical vapour deposition (CVD), reactive sputtering, hot spraying and reactive evaporation [6]. TCOs are widely used in applications such as flat-panel displays, architectural heat reflecting coatings and PV panels [7], where they transmit visible light and reflect infrared radiation. For TPV, the plasma wavelength needs to be shifted into the infrared. TCOs have been considered on the heat shield [47, 48], as PV cell front surface filter (FSF) [49] and as an integral part of III-V semiconductor cells (e.g., InGaAs) [7, 50].

The bandgap of these materials is typical around or above 3 eV and this value corresponds to a wavelength of about 0.4 μm. Hence, these materials are transparent in the visible and part of the infrared spectral region. Doping of these semiconductors causes a high electrical conductivity. This in turn leads to reflection of infrared radiation beyond the plasma frequency.

The reflectance versus wavelength behaviour can been modelled using the classical Drude theory. Equation (3.1) defines the plasma frequency ν_p, where N is the number of charge carries, e_o the elementary charge, ε_0 the vacuum dielectric constant, ε_b the dielectric constant associated with bound carriers at very high frequency, m^* the effective mass and Y the relaxation frequency of the semiconductor. The cut-off wavelength, or the plasma wavelength, can be shifted by a change of the number of charge carries N [9, 10, 51].

$$\nu_p = \sqrt{\frac{N \cdot e_0^2}{\varepsilon_b \varepsilon_0 m^*} - Y^2} \tag{3.1}$$

$$Q = \frac{N\mu^2 m^*}{\varepsilon_b \varepsilon_0} - 1 \tag{3.2}$$

In order to get a steep transition from transmittance to reflectance the quality factor Q, given by Eq. (3.2) needs to be maximised. An important parameter is the electron mobility μ that needs to be maximised [9, 10].

The advantages of TCOs are potentially low costs and broad reflectance in the mid and far infrared. One major disadvantage of TCOs is that this filter has significant absorption that is strongest near to the plasma frequency. For example, cadmium stannate filters demonstrated mid and far infrared reflectance over 90% with an absorption value of around 10% [52]. Another drawback is the slop characteristic from transmission to reflection. In other words, TCO in the infrared do not show a very sharp cut-off at the plasma frequency [5]. In order to improve the performance, a high electron mobility μ is required. Further improvements of TCOs may or may not allow reaching the high mid and far infrared reflectance of FSS filters [52]. One important development is the combination of TCO filters with a wide spectral range and all-dielectric filters with a small spectral range but sharper cut-off characteristic. Section 3.7 discusses this tandem filter type further.

3.5 All-Dielectric Filters

All-dielectric filters, also called interference or dichroic filters [4], consist of multiple thin layers of material having different refractive indices [7, 45]. All-dielectric filters are well understood and they can be flexibly designed as bandpass or band-rejection filter [7]. The ratio of high to low refractive index should be maximised. This allows a filter design with fewer layers, less total material, and thus less total absorption, and lower fabrication cost [53]. Software packages allow optimisation of the transmission and reflection versus wavelength characteristics of the filters for given optical material constants (e.g., refractive index versus wavelength) [7].

Materials with high refractive indices include zinc selenide ($n_{ZnSe} = 2.4$), zinc sulphide ($n_{ZnS} = 2.3$) and silicon. Magnesium fluoride ($n_{MgF2} = 1.4$) is a typical material with a low index value [2, 5, 7, 45]. For TPV conversion, the thermal stability of these film materials can be an issue. Also, there are efforts to increase the refractive index contrast. Hence, alternative filter materials have been examined. These are the high index materials antimony sulphide ($n_{Sb2S3} = 2.8$), gallium telluride ($n_{GaTe} = 3.0$) and antimony selenide ($n_{Sb2Se3} = 3.4$), where the maximum temperature of Sb_2Se_3 was limited to 90°C. The alternative low index material was yttrium fluoride ($n_{YF3} = 1.5$) [53]. Developments using silicon (high index) and silicon dioxide (low index) did not have the required performance [5].

All-dielectric filters are advantageous in terms of their low absorption. On the other hand, these filters have two major disadvantages when they are applied for spectral control in a TPV system. First, these filters are typically optimised for smaller wavelength ranges. Also combining more than one all-dielectric filter cannot overcome the limited spectral range due to interference between the filters [4, 45]. The TPV requirements of out-of-band reflectivity over a wide wavelength range, a high in-band transmissivity and sharp transition behaviour can only be met using a large number of dielectric layers. Challenges of filter designs with a large number of layers include high costs, the control of film thickness and adherence of the layers [2, 7]. Second, it is well known that the transmission and reflection versus wavelength of dielectric filters are a function of the incident angle. Hence, optical systems have been designed to bundle the radiation so that the incident radiation at the filter has small zenith angles (see Sect. 6.3.3) [54].

3.6 Metal-Dielectric Filters

The basic theoretical description of the metal-dielectric filter (also called induced transmission filter) was published by Berning and Turner as early as 1957 [6, 55]. The filter consists of a thin metal layer sandwiched between one or more dielectric layer on each side. Typical metals included gold and silver with dielectric materials such as ZnS, ZnSe, MgF_2, ZnO and TiO_2 [2, 4, 6, 19, 45, 56–58]. For the spectral range in TPV conversion, the metal layer thickness is typically in a range from 10 to 30 nm.

Metal films have the advantage of high reflectivity in the mid and far infrared wavelength range. The major disadvantage of the filter is the parasitic absorption in the metal layer [2, 4, 6, 45]. Similar to all-dielectric filters, it is necessary to adjust the thickness of all layers (dielectric and metal layer) to optimise the overall performance of the filter.

3.7 Combined Dielectric-TCO Filters

Another concept is a *tandem filter* that consists of an all-dielectric and a TCO filter [2]. This concept was introduced in the early 1990s by researchers at Knolls Atomic Power with a proposal in the first TPV Conference using indium tin oxide as a plasma filter [1, 2, 5, 8]. The filter sits usually on the front surface of the PV cell. The filter can be glued with an optical adhesive (epoxy) onto the PV cell. This arrangement allows cooling of the filter via the cell. The dielectric filter acts as the first surface mirror for medium wavelengths (e.g., 2–6 μm). Long wavelength radiation (e.g., larger than 6 μm) is reflected by the plasma filter that is located between the dielectric filter and the cell [2, 53, 59].

These tandem filters achieved a high performance and there is a good agreement of the performance by theoretical modelling and practical measurements [1, 5, 44, 60, 61]. For example, a tandem filter, using multiple layers of Sb_2Se_3/YF_3 as dielectric filter and highly n-doped InPAs as plasma filter, was glued on an InGaAs MIM cell with a bandgap of 0.6 eV. The radiant heat transfer efficiency of the module at 25°C was 23.6%. This system used a $5.4 \times 5.4\ cm^2$ etched SiC graybody radiator at 1,039°C placed 2 mm away from the 2×2 cm cell module using a vacuum gap as insulation. The electrical power density was about 0.8 W/cm^2. It should be pointed out that the MIM using a back surface reflector without the tandem filter had also a high performance (20.6%, 0.9 W/cm^2) [59].

As pointed out in the discussion of dielectric filters, tandem filters need to overcome some fabrication issues due to the large number of layers, such as control of thickness and adherence of the layers, for large-scale usage. It can be pointed out that the tandem filter approach has some flexibility, since the spectral range of the two filters are not fixed but could be altered. For example, if there would be technological progress in the performance of one filter (e.g., low absorption of plasma filter), the contribution of this filter could be extended by a wider spectral range.

3.8 Other Filter Concepts

In the late 1970s and early 1980s the combination of *dielectric filters* and PV cell *BSR* was researched. This approach was not very successful, since the long wavelength reflectivity of both the dielectric filter and the BSR were rather poor [5].

Rugate filters are based on interference coatings with a continuously varying refractive index. This filter was proposed as FSF in full-spectrum cavity solar converters [62, 63].

Spectral splitters apply different principles such as reflection/transmission [62, 63], refraction (e.g., prism) and holography. Only TPV work on the latter could be identified [64, 65]. There are also spectral splitting systems for solar conversion. Imenes and Mills wrote a review article of spectral beam splitting [65]. Applications include hybrid lighting and full-spectrum cavity converters [62, 63, 66]. Such systems can be considered as non-TPV type because of the absence of a radiator. Although direct solar spectral splitting systems may use the same low bandgap PV cells as TPV systems.

At the Paul Scherrer Institute (PSI) a 5 mm thick *water* layer held by two concentrical SiO_2 tubes has been considered as a long wavelength absorption filter (above 1.4 μm) [48, 67]. Advantages of this design include simplicity, high absorption of mid and far infrared radiation and the use of the hot water for a CHP system. Disadvantages include reduced in-band transmission (or power density) and reduction of electrical conversion efficiency due to the absorbed and "lost" heat in the water.

The *inverse filter* concept, in which undesirable long wavelength radiation (out-of-band) is transmitted and convertible short wavelength radiation (in-band) is reflected, has also been considered for TPV [45]. Such filters are known as cold mirrors and are used for example in dichroic lamps. It has been pointed out that TPV systems utilising cold mirrors would require a complex design with at least one additional mirror [45].

3.9 Summary

In early TPV systems filters were placed within the cavity (e.g., onto the heat shield) and this resulted in the overheating of the filter. In the course of the development of various systems, there was a tendency to place the filter closer and closer to the PV cell to minimise the temperature of the filter. Thermal stability issues and the temperature dependent performance of filters resulted in the logical step to locate the filter directly on the cell (front-surface filter) and this filter design can be considered as state of the art. This approach allows cooling of the filter via the cell, since all filters will show some absorption that causes heating of the filter. The filter can be either mechanically mounted onto the PV cell, using for example optical glue, or the filter can be an integral part of the cell design. For example, Abbott examined metal-dielectric and plasma filters deposited directly on the cell surface [58].

In the previous sections the weaknesses and strengths of the different filter types were discussed. All-dielectric filter can have low absorption but only in a narrow spectral range, whereas TCO and metal-dielectric filters have a wide spectral range but show typically significant absorption and no spectrally sharp filter characteristic. Hence, the combination of different filter types can lead to a better overall performance. In particular the tandem dielectric-TCO filter showed a high performance. Another promising filter concept is the FFS filter with a very high mid and long wavelength reflectivity. Both, tandem dielectric-TCO and FFS filters demonstrate that high performance filters can be designed. Also spectral control within the cell, e.g. BSR in a MIM module, achieved a suitable performance. However, currently there is still a lack of economic and large-scale filter solutions for spectral control. This situation can be seen as one major cause that only few complete efficient TPV cavities have been realised.

References

1. Good BS, Chubb DL, Lowe RA (1997) Comparison of selective emitter and filter thermophotovoltaic systems, Proceeding of the 2nd. NREL Conference on Thermophotovoltaic Generation of Electricity, Colorado, Springs, 16–20 July 1995. American Institute of Physics, pp 16–34
2. Chubb D (2007) Fundamentals of Thermophotovoltaic Energy Conversion. Elsevier Science, Amsterdam
3. Chubb D, Nelson R (1995) Workshop 3: Emission & spectral control. Proceeding of the 1st. NREL Conference on thermophotovoltaic generation of electricity, Copper Mountain, Colorado, 24–28 July 1994. American Institute of Physics, pp 13–16
4. Gruenbaum PE, Kuryla MS, Sundaram VS (1995) Technical and economic issues for gallium antimonide based thermophotovoltaic systems. Proceeding of the 1st. NREL Conference on thermophotovoltaic generation of electricity, Copper Mountain, Colorado, 24–28 July 1994. American Institute of Physics, pp 357–367
5. Horne WE, Morgan MD, Sundaram VS (1996) IR filters for TPV converter modules. Proceeding of the 2nd. NREL Conference on thermophotovoltaic generation of electricity, Colorado Springs, 16–20 July 1995. American Institute of Physics, pp 35–51
6. Köstlin H (1982) Application of Thin Semiconductor and Metal Films in Energy Technology. Festkörperprobleme 22:229–254
7. Coutts TJ (1999) A review of progress in thermophotovoltaic generation of electricity. Renew Sustain Energy Rev 3(2–3):77–184
8. Baldasaro PF, Brown EJ, Depoy DM, Campbell BC, Parrington JR (1995) Experimental assessment of low temperature voltaic energy conversion. Proceeding of the 1. NREL Conference on thermophotovoltaic generation of electricity, Copper Mountain, Colorado, 24–28 July 1994. American Institute of Physics, pp 29–43
9. Silberglitt R, Le HK (1991) Materials for Solar Collector Concepts and Designs. In: de Winter F (ed) Solar Collectors, Energy Storage and Materials, Chap 21. MIT Press, Cambridge, MA
10. Lampert CM (1991) Theory and Modeling of Solar Materials. In: de Winter F (ed) Solar Collectors, Energy Storage and Materials, Chap 22. MIT Press, Cambridge, MA
11. van den Hoek WJ, Jack AG, Luijks GMJF (2005) Lamps, in Ullmanns Encyclopedia of Industrial Chemistry. Wiley, London
12. Goldstein MK, DeShazer LG, Kushch AS, Skinner SM (1997) Superemissive light pipe for TPV applications, Proceeding of the 3rd. NREL Conference on thermophotovoltaic generation of electricity, Denver, Colorado, 18–21 May 1997. American Institute of Physics, pp 315–326
13. Tropf WJ, Thomas ME, Harris TJ (1995) Properties of Crystals and Glasses. In: Bass M (ed) Handbook of Optics, Chap 33. McGraw-Hill, New York

14. Chubb DL (2007) Light Pipe Thermophotovoltaics (LTPV), Proceeding of the 7th. World Conference on thermophotovoltaic generation of electricity, Madrid, 25–27 Sept 2006. American Institute of Physics, pp 297–316
15. Lide DR (2008–2009) Physical Constants of Inorganic Compounds, in CRC Handbook chemistry and physics, 89th edn. CRC Press
16. (2009) Global Vacuum Product Guide, Section 17 - Technical Information, 9. Edition, Kurt J. Lesker Company [Online] Available at: http://www.lesker.com/. Accessed 28 Apr 2010
17. Browder JS, Ballard SS, Klocek P (1991) Physical property comparison of infrared optical materials. In: Klocek P (ed) Handbook of infrared optical materials. Marcel Dekker Inc, New York
18. Pierce DE, Guazzoni G (1999) High temperature optical properties of thermophotovoltaic emitter components. Proceeding of the 4th NREL Conference on thermophotovoltaic generation of electricity, Denver, Colorado, 11–14 Oct 1998. American Institute of Physics, pp 177–190
19. Guazzoni GE, Rose MF (1996) Extended use of photovoltaic solar panels. Proceeding of the 2nd. NREL Conference on thermophotovoltaic generation of electricity, Colorado Springs, 16–20 July 1995. American Institute of Physics, pp 162-176
20. Palfinger G, Bitnar B, Durisch W, Mayor J-C, Grützmacher D, Gobrecht J (2003) Cost estimate of electricity produced by TPV. Semicond Sci Technol 18:254–261
21. Fraas L, Samaras J, Han-Xiang Huang Seal M, West E (1999) Development status on a TPV cylinder for combined heat and electric power for the home. Proceeding of the 4th NREL Conference on thermophotovoltaic generation of electricity, Denver, Colorado, 11–14 Oct 1998. American Institute of Physics, pp 371–383
22. De Jong BHWS, Beerkens RGC, van Nijnatten PA (2005) Glass, in Ullmanns Encyclopedia of Industrial Chemistry. Wiley-VCH Verlag GmbH & Co. KGaA, Weinheim.
23. Doremus RH (1973) Glass Science. Wiley, New York, pp 74–97
24. Hetherington G, Jack KH (1963) Fused Quartz and Fused Silica, TSL Wallsend Northumberland England, translated and reprinted from Ullmanns Encyklopädie der technischen Chemie, 14, 3rd edn. Urban & Schwarzberg, München Berlin, pp 511–524
25. Kunc T, Lallemand M, Saulnier JB (1984) Some new developments on coupled radiative-conductive heat transfer in glasses–experiments and model. Int J Heat Mass Transfer 27: 2307–2319
26. Edwards OJ (1966) Optical transmittance of fused silica at elevated temperatures. J Opt Soc Am 56:1314–1319
27. Philipp HR (1985) Silicon dioxide (SiO2) (Glass). In: Palik ED (ed) Handbook of optical constants of solids. Academic Press, London, pp 749–763
28. Barsoum M (1997) Fundamentals of Ceramics. McGraw-Hill, New York, pp 611–649
29. White DC, Hottel HC (1995) Important factors in determining the efficiency of TPV systems. Proceeding of the 1st. NREL Conference on Thermophotovoltaic Generation of Electricity. Copper Mountain, Colorado, 24–28 July 1994. American Institute of Physics, pp 425–454
30. Fraas LM, Ferguson L, McCoy LG, Pernisz UC (1996) SiC IR Emitter Design for Thermophotovoltaic Generators. Proceeding of the 2nd NREL Conference on thermophotovoltaic generation of electricity, Colorado Springs, 16–20 July 1995. American Institute of Physics, pp 488–494
31. Hanamura K, Kumano T (2003) Thermophotovoltaic power generation by super-adiabatic combustion in porous quartz glass. Proceeding of the 5th Conference on thermophotovoltaic generation of electricity, Rome, 16–19 Sept 2002. American Institute of Physics, pp 111–120
32. Hanamura K, Kumano T (2004) TPV power generation system using super-adiabatic combustion in porous quartz glass. Proceeding of the 6th International Conference on thermophotovoltaic generation of electricity. Freiburg, Germany, 14–16 June 2004. American Institute of Physics, pp 88–95
33. Kumano T, Hanamura K (2004) Spectral control of transmission of diffuse irradiation using piled AR coated quartz glass filters. Proceeding of the 6th International Conference on thermophotovoltaic generation of electricity, Freiburg, Germany, 14–16 June 2004. American Institute of Physics, pp 230–236

34. Bauer T, Forbes I, Penlington R, Pearsall N (2005) Heat transfer modelling in thermophotovoltaic cavities using glass media. Sol Energy Mater Sol Cells 88(3):257–268
35. Vardaxoglou JC (1997) Introduction to frequency selective surfaces. In: Vardaxoglou JC (ed) Frequency Selective Surfaces: Analysis and Design, Chap. 1. Wiley, New York, pp 1–13
36. Jefimovs K, Vallius T, Kettunen V, Kuittinen M, Turunen J, Vahimaa P, Kaipiainen M, Nenonen S (2004) Inductive grid filters for rejection of infrared radiation. J Mod Opt 51:1651–1661
37. Reed JA (1997) Frequency Selective Surfaces with Multiple Periodic Elements. Ph.D thesis. University of Texas, Dallas
38. Spector SJ, Astolfi DK, Doran SP, Lyszczarz TM, Raynolds JE (2001) Infrared Frequency Selective Surfaces Fabricated using Optical Lithography and Phase-Shift Masks, Report, Lockheed Martin Corporation, US, LM-01K062
39. Horne WE, Morgan MD, Sundaram VS, Butcher T (2003) 500 watt diesel fueled TPV portable power supply. Proceeding of the 5th Conference on thermophotovoltaic generation of electricity, Rome, Italy, 16–19 Sept 2002. American Institute of Physics, pp 91–100
40. Horne E (2002) Hybrid Thermophotovoltaic Power Systems. EDTEK Inc. US Consultant Report, P500-02-048F
41. Horne WE, Morgan MD (1995) Filter array for modifying radiant thermal energy. EDTEK Inc., US Patent 5,611,870
42. Horne W, Morgan M, Horne W, Sundaram V (2004) Frequency selective surface bandpass filters applied to thermophotovoltaic applications. Proceeding of the 6th International Conference on thermophotovoltaic generation of electricity, Freiburg, Germany, 14–16 June 2004. American Institute of Physics, pp 189–197
43. Schock A, Kumar V (1995) Radioisotope thermophotovoltaic system design and its application to an illustrative space mission. Proceeding of the 1st NREL Conference on thermophotovoltaic generation of electricity. Copper Mountain, Colorado, 24–28 July 1994. American Institute of Physics, pp 139–152
44. Kristensen RT, Beausang JF, DePoy DM (2004) Frequency selective surfaces as near-infrared electromagnetic filters for thermophotovoltaic spectral control. Appl Phys 95:4845–4851
45. Höfler H (1984) Thermophotovoltaische Konversion der Sonnenenergie (in German), Doctoral thesis. Universität Karlsruhe (TH)
46. Ehsani H, Bath I, Borrego J, Gutmann R, Brown E, Dzeindziel R, Freeman M, Choudhury N (1996) Characteristics of degenerately doped silicon for spectral control in thermophotovoltaic systems. Proceeding of the 2nd NREL Conference on thermophotovoltaic generation of electricity, Colorado Springs, 16–20 July 1995. American Institute of Physics, pp 312–328
47. Anna Selvan JA, Grützmacher D, Hadorn M, Bitnar B, Durisch W, Stutz S, Neiger T, Gobrecht J (2000) Tuneable plasma filters for TPV systems using transparent conducting oxides of tin doped indium oxide and Al doped zinc oxide. Proceeding of the 16th. European Photovoltaic Solar Energy Conference, James and James Science, pp 187–190
48. Bitnar B, Durisch W, Grützmacher D, Mayor J-C, von Roth F, Anna Selvan JA, Sigg H, Gobrecht J (2000) Photovoltaic cells for a thermophotovoltaic system with a selective emitter. Proceeding of the 16th European Photovoltaic Solar Energy Conference, James and James Science, pp 191–194
49. Murthy SD, Langlois E, Bath I, Gutmann R, Brown E, Dzeindziel R, Freeman M, Choudhury N (1996) Characteristics of indium oxide plasma filters deposited by atmospheric pressure CVD. Proceeding of the 2nd NREL Conference on thermophotovoltaic generation of electricity, Colorado Springs, 16–20 July 1995. American Institute of Physics, pp 290–311
50. Charache GW, DePoy DM, Raynolds JE, Baldasaro PF, Miyano KE, Holden T, Pollak FH, Sharps PR, Timmons ML, Geller CB, Mannstadt W, Asahi R, Freeman AJ, Wolf W (1999) Moss–Burstein and plasma reflection characteristics of heavily doped n-type InxGa1-xAs and InPyAs1-y. J Appl Phys 86:452–458
51. Zenker M (2001) Thermophotovoltaische Konversion von Verbrennungswaerme (in German). Doctoral thesis, Albert-Ludwigs-Universität Freiburg im Breisgau

52. Wu X, Mulligan WP, Webb JD, Coutts TJ (1996) TPV plasma filters based on cadmium stannate. Proceeding of the 2nd NREL Conference on Thermophotovoltaic Generation of Electricity, Colorado Springs, 16–20 July 1995. American Institute of Physics, pp 329–338
53. Rahmlow TD, DePoy DM, Fourspring PM, Ehsani H, Lazo-Wasem JE, Gratrix EJ (2007) Development of Front Surface, Spectral Control Filters with Greater Temperature Stability for Thermophotovoltaic Energy Conversion. Proceeding of the 7th World Conference on thermophotovoltaic generation of electricity, Madrid, 25–27 Sept 2006. American Institute of Physics, pp 59–67
54. Lindberg E (2002) TPV Optics Studies - On the Use of Non-imaging Optics for Improvement of Edge Filter Performance in Thermophotovoltaic Applications. Doctoral thesis, Swedish University of Agricultural Science, Uppsala
55. Berning PH, Turner AF (1957) Induced Transmission in Absorbing Films Applied to Band Pass Filter Design. J Opt Soc Am 47:230–239
56. Höfler H, Paul HJ, Ruppel W, Würfel P (1983) Intereference filter for thermophotovoltaic solar energy conversion. Sol Cells 10(3):273–286
57. Demichelis F, Minetti-Mezzetti E, Agnello M, Perotto V (1982) Bandpass filters for thermophotovoltaic conversion systems. Sol Cells 5:135–141
58. Abbott P, Bett AW (2004) Cell-mounted spectral filters for thermophotovoltaic applications. Proceeding of the 6th International Conference on thermophotovoltaic generation of electricity, Freiburg, Germany, 14–16 June 2004. American Institute of Physics, pp 244–251
59. Wernsman B, Siergiej RR, Link SD, Mahorter RG, Palmisiano MN, Wehrer RJ, Schultz RW, Schmuck GP, Messham RL, Murray S, Murray CS, Newman F, Taylor D, DePoy DM, Rahmlow T (2004) Greater than 20% radiant heat conversion efficiency of a thermophotovoltaic radiator/module system using reflective spectral control. Trans Electron Dev 51(3):512–515
60. Fourspring PM, DePoy DM, Beausang JF, Gratrix EJ, Kristensen RT, Rahmlow TD, Talamo PJ, Lazo-Wasem JE, Wernsman B (2004) Thermophotovoltaic spectral control. Proceeding of the 6th International Conference on thermophotovoltaic generation of electricity, Freiburg, Germany, 14–16 June 2004. American Institute of Physics, pp 171–179
61. Rahmlow TD, Lazo-Wasem JE, Gratrix EJ, Fourspring PM, DePoy DM (2004) New performance levels for TPV front surface filters. Proceeding of the 6th International Conference on thermophotovoltaic generation of electricity, Freiburg, Germany, 14–16 June 2004. American Institute of Physics, pp 180–188
62. Ortabasi U, Bovard BG (2003) Rugate technology for thermophotovoltaic (TPV) applications a new approach to near perfect filter performance. Proceeding of the 5th Conference on thermophotovoltaic generation of electricity, Rome, 16–19 Sept 2002. American Institute of Physics, pp 249–258
63. Ortabasi U, Friedman HW (2004) PowerSphere: A novel photovoltaic cavity converter using low bandgap TPV cells for efficient conversion of high power laser beams to electricity. Proceeding of the 6th International Conference on thermophotovoltaic generation of electricity, Freiburg, Germany, 14–16 June 2004. American Institute of Physics, pp 142–152
64. Regan TM, Martin JG, Riccobono J (1995) TPV conversion of nuclear energy for space applications. Proceeding of the 1st. NREL Conference. on thermophotovoltaic generation of electricity. Copper Mountain, Colorado, 24–28 July 1994. American Institute of Physics, pp 322–330
65. Imenes AG, Mills DR (2004) Spectral beam splitting technology for increased conversion efficiency in solar concentrating systems: a review. Sol Energy Mater and Sol Cells 84:19–69
66. Fraas LM, Daniels WE, Muhs J (2001) Infrared Photovoltaics for Combined Solar Lighting and Electricity for Buildings. Proceeding of the 17th. European Photovoltaic Solar Energy Conference, Munich, 22–26 Oct 2001. WIP
67. Durisch W, Grob B, Mayor JC, Panitz JC, Rosselet A (1999) Interfacing a small thermophotovoltaic generator to the grid, 4th NREL Conference on thermophotovoltaic generation of electricity, Denver, Colorado, 11–14 Oct 1998, American Institute of Physics, pp 403–414

Chapter 4
Photovoltaic Cells

4.1 Introduction

The major PV cell research interests are the development of efficient low bandgap PV cells, the incorporation of spectral control (filters and reflectors), multijunction cells and improvement of the economics (e.g., alternative substrates). Different authors reviewed PV semiconductors and theoretical PV cell aspects for TPV conversion. These authors include Coutts [1, 2], Andreev [3], Bhat et al. [4], Iles [5], Woolf [6, 7] and Chubb [8].

This chapter has the following structure. Section 4.2 covers the PV cell theory with basic aspects about the current-voltage characteristics of a PV cell and the assessment of the cell performance via partial PV cell efficiencies. Section 4.3 discusses fabrication technologies for PV cells. Section 4.4 shows the options to include spectral control as an integral part of the PV cell design. The performance of PV cells based on group IV (Sect. 4.5) and group III-V semiconductors (Sect. 4.6) is assessed by the previously defined partial cell efficiency. Section 4.7 discusses alternative PV cell materials, as well as other aspects and components related to the PV cell.

4.2 PV Cell Theory

The less common approach presented in the section is based on the assessment of the PV cell performance by partial cell efficiencies. The efficiencies are the voltage factor η_{OC}, the collection efficiency (also called mean quantum efficiency) η_{QE}, the fill factor η_{FF}, the ultimate efficiency η_{UE} and the PV cell array efficiency η_{Array} Eq. 4.1 [4, 9–15].

$$\eta_{PV} = \frac{P_{el}}{P_{PV}} = \eta_{OC} \cdot \eta_{QE} \cdot \eta_{FF} \cdot \eta_{UE} \cdot \eta_{Array} \tag{4.1}$$

T. Bauer, *Thermophotovoltaics*, Green Energy and Technology,
DOI: 10.1007/978-3-642-19965-3_4,

The first three efficiencies (η_{OC}, η_{QE} and η_{FF}) are mainly related to the characteristics of the PV cell and an in depth discussion follows in this section. The *ultimate efficiency* η_{UE} describes losses due to the spectral mismatch of the absorbed illumination spectrum in relation to the PV cell bandgap. The ultimate efficiency η_{UE} describes how the absorbed radiation $P_{PV(v)}$ matches the PV cell bandgap energy hv_g. There are two loss mechanisms. The first of these is due to photons with energies $hv < hv_g$ which are not converted and this mechanism is named *free carrier heating* . In other words, a photon with energy less than the bandgap energy hv_g makes no contribution to the electrical cell output. The second mechanism is due to photons with energies $hv > hv_g$. These photons contribute to the energy output hv_g and the excess energy ($hv–hv_g$) is lost as heat. This second mechanism is named *hot carrier heating* [10, 16, 17]. Optimisation of the ultimate efficiency is an essential aspect of a TPV system including for example a filter for spectral control. Hence, it cannot be considered a sole cell issue. Section 6.2 discusses the dependency of the ultimate efficiency for different illumination spectra (suppression of radiation above and below the bandgap).

The discussed efficiencies all relate to the single cell performance under ideal conditions. The efficiency η_{Array} includes losses due to multiple cell connections. PV cells connected in series and parallel and built into TPV systems have generally lower efficiencies than a single cell because of the following reasons:

- Distribution of different individual cell properties, so that each cell cannot operate at its own maximum power point. Also a group of cells and not only the best performing cells are used.
- Cells within a group of cells operating at different radiation densities; non-uniform absorption of radiation due to the spatial and angular distribution of the radiation in a TPV cavity.
- Packing of the PV cells with uncovered areas (for TPV, the packing factor has been optimised using shingling [18]).
- Conduction and convection heat transfer to the PV cell; by definition: the absorbed radiation power $P_{PV(v)}$ does not include conduction and convection losses, whereas the absorbed radiation power $P_{PV\#(v)}$ includes these losses (compare energy balance in Fig. 1.2).

For TPV conversion, it should be pointed out that many of the discussed negative factors (e.g., non-uniform cell temperature and incident angle of the radiation) are temporally constant compared to solar PV. Hence, non-uniformities can be analysed and then minimised due to a suitable system design.

4.2.1 Current-Voltage Characteristics

The conventional photovoltaic cell has a p-n junction and a single bandgap. Figure 4.1 on the right hand side shows the equivalent circuit of an ideal photovoltaic cell. More realistic models include series and parallel resistances, as well as

$$J = J_s\left(e^{\frac{e_0V}{kT_{cell}}} - 1\right) - J_{ph}$$

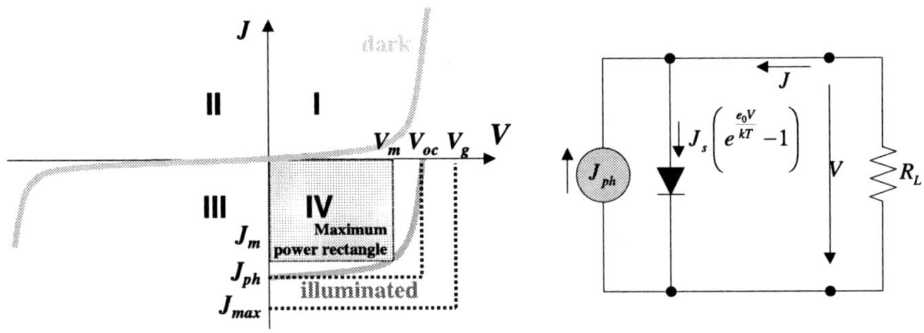

Fig. 4.1 Schematic of the current-voltage curve (*left*) and the idealised equivalent circuit (*right*)

other diode parameters, such as additional diodes or a diode ideality factor [19]. For simplicity, the following discussion focuses on the presented ideal photovoltaic cell model. A constant current source is in parallel with a diode. In the dark, there is no current generated by photons and the photocurrent density J_{ph} is zero. The current-voltage characteristic can be described by the Shockley diode equation Eq. 4.2. Figure 4.1 on the left hand side shows schematically the current-voltage characteristic of the dark cell with a photocurrent $J_{\mathrm{ph}} = 0$. For the illuminated cell, ideally each photon above the bandgap $h\nu_{\mathrm{g}}$ contributes to one elementary charge e_0. The generated photocurrent per area is called photocurrent density J_{ph}. When a photovoltaic cell is illuminated, the current-voltage curve is shifted to the fourth quadrant (Fig. 4.1 left) and electrical power can be extracted from the cell and utilised in the external load R_{L}. There is a specific current density J_{m} and voltage V_{m} where the maximum power output, or maximum area under the curve, is obtained (Fig. 4.1, left).

$$J = J_{\mathrm{s}}\left(e^{\frac{e_0 V}{kT_{\mathrm{cell}}}} - 1\right) - J_{\mathrm{ph}} \tag{4.2}$$

4.2.2 Dark Saturation Current Density

The diode equation includes the dark saturation current density J_{s} Eq. 4.2. Equation 4.3 defines the current J_{s}, where e_0 is the elementary charge, n_{i} the intrinsic carrier concentration, D is the diffusion coefficient, L is the diffusion length and N_{A} and N_{D} are the acceptor and donor doping concentrations, respectively. The subscripts e and h refer to electron and holes as minority carriers [16, 19]. It can be seen that several material parameters are required in order to calculate the dark saturation current density. Hence, modelling of the impact of the PV cell bandgap is usually not carried out by individually calculated J_s-values for each cell. Typically such modelling utilises the thermodynamic limit of J_s or an empirical equation depending on the bandgap energy as discussed in the following.

$$J_s = e_0 n_i^2 \left(\frac{D_e}{N_A L_e} + \frac{D_h}{N_D L_h} \right) \tag{4.3}$$

The dark saturation current density J_s has a thermodynamic lower limit because the cell not only absorbs radiation but also emits radiation. This theoretical radiative limit is given by Eq. 4.4. Equation 4.4 is valid for $h\nu_g > 0.4$ eV with an error of less than 2% in J_s. This equation requires also assumptions of the Étendue. Equation 4.4 assumes an Étendue proposed by Henry with a radiative limit without photon recycling using a typical semiconductor refractive index (here n = 3.6) and with photon recycling (n = 0) [20]. An in depth analysis of the radiative limit and other Étendue types can be found elsewhere [21]. Frequently, an approximate solution that neglects the kT_{cell}-terms in Eq. 4.4 in brackets is used Eq. 4.5 [9, 22]. The approximation is only valid if $h\nu_g \gg kT_{cell}$. The approximation results in an error of the J_s value of about 10%, if a bandgap of 0.5 eV and a cell temperature of 25°C are assumed [21].

$$J_s = \frac{2e_0}{h^3 c_0^2} \underbrace{\pi \left(n^2 + 1\right)}_{\text{Étendue}} kT_{cell} \left(2k^2 T_{cell}^2 + 2kT_{cell} h\nu_g + \left(h\nu_g\right)^2 \right) e^{-\frac{h\nu_g}{kT_{cell}}} \tag{4.4}$$

$$J_s \approx \frac{2e_0}{h^3 c_0^2} \pi \left(n^2 + 1\right) kT_{cell} \left(h\nu_g\right)^2 e^{-\frac{h\nu_g}{kT_{cell}}} \tag{4.5}$$

Frequently, models of J_s depending on $h\nu_g$ use an approximation in the form as defined in Eq. 4.6 [5–7] and Eq. 4.8 with a constant β [23]. Sze reported Eq. 4.7 for high radiation densities [16]. Coutts reports about an empirical correlation using Eqs. 4.8 and 4.9, where the latter depends on the bandgap energy. Wanlass fitted J_s values for a wide range of semiconductors in order to determine the parameters in Eq. 4.9 [1, 2, 24]. Mauk cites some further literature for the modelling of J_s [25].

$$J_s \approx \beta e^{-\frac{h\nu_g}{kT_{cell}}} \tag{4.6}$$

$$J_s \approx \beta T_{cell}^{\frac{3}{2}} e^{-\frac{h\nu_g}{2kT_{cell}}} \tag{4.7}$$

$$J_s \approx \beta_{(h\nu_g)} T_{cell}^3 e^{-\frac{h\nu_g}{kT_{cell}}} \tag{4.8}$$

$$\beta_{(h\nu_g)\text{Wanlass}} = 3.165 \cdot 10^{-4} \frac{A}{cm^2 K^3} e^{2.91 h\nu_g} \tag{4.9}$$

The performance of different PV cells can be compared using the described J_s-models Eq. 4.4–4.9 [26]. The J_s-value of a characterised PV cell can be

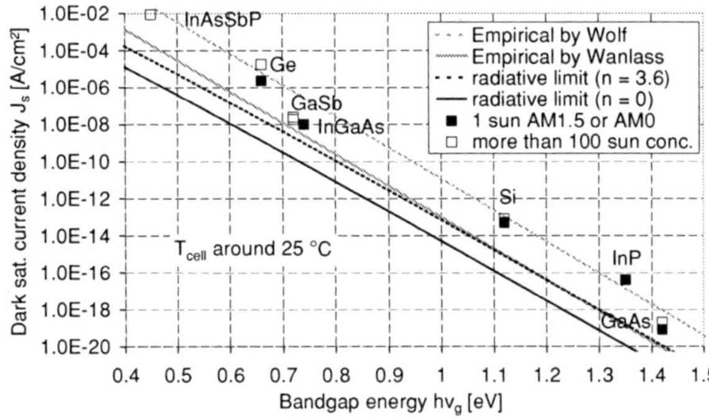

Fig. 4.2 Dark saturation current density vs. bandgap for different concentration levels; the figure was generated using literature data [27–32]

determined by Eq. 4.10 using the open circuit voltage, the photocurrent and the cell temperature. This equation is based on the rearrangement of the ideal diode equation with $J = 0$ (Eq. 4.2). Figure 4.2 shows a comparison of models and calculated J_s values of different cells. It can be seen that high-performance III-V semiconductors, such as GaSb, InGaAs and GaAs are in good approximation described by the model with empirical values from Wanlass [1, 2, 24]. The simple model by Wolf fits better the group of IV semiconductors, namely silicon and germanium [6, 7].

$$J_s = \frac{J_{ph}}{e^{\frac{e_0 V_{oc}}{kT_{cell}}} - 1} \tag{4.10}$$

4.2.3 Collection Efficiency

For an ideal cell the photocurrent density, or short circuit current density, is given by Eq. 4.11. The quantity of electrons and holes collected per photon absorbed is called the quantum efficiency. The quantum efficiency can include reflection losses (external quantum efficiency) or exclude the reflected radiation of the cell surface $R_{(v)}$ (internal quantum efficiency). The quantum efficiency depends on the frequency, or wavelength, and is an inherent property to the device independent of the radiation source [27, 33]. An increase in cell temperature leads to a reduction of the PV cell bandgap energy hv_g. As a result, the number of integrated photons increases and the photocurrent density J_{ph} rises slightly. Equations 4.12 and 4.13 define the spectral response (in A/W) utilised for cell characterisation, where $I_{(v)}$ is the radiation intensity (in W/m^2) [16, 34].

$$J_{ph} = e_0 \int_{v_g}^{\infty} \underbrace{\left(1 - R_{(v)}\right) \eta_{QE,int(v)}}_{\eta_{QE,ext(v)}} n_{ph(v)} \mathrm{d}v \tag{4.11}$$

$$SR_{(\nu)} = \frac{J_{ph(\nu)}}{I_{(\nu)}} \tag{4.12}$$

$$\eta_{QE,ext(\nu)} = SR_{(\nu)} \frac{h\nu}{e_0} \tag{4.13}$$

To obtain the collection efficiency, (Eq. 4.11) can be rearranged. The maximum photocurrent density J_{max} is given by the number of photons with energies above the bandgap energy $h\nu_g$ presented in the radiation spectrum multiplied with the elementary charge e_0 (Eq. 4.14). The number of photons is also given by the absorbed radiation power of the cell divided by the product of cell area and photon energy $h\nu$. It needs to be considered that the $P_{PV(\nu)}$ does not include parasitic conduction and convection losses. Also, radiation from the emitter can be lost in the cavity due to a view factor less than unity. Hence, the radiation absorbed by the cell is not necessarily the radiation emitted by the radiator (compare energy balance in Sect. 1.5). The actual photocurrent density J_{ph} is smaller than J_{max} and this reduction can be described by the *collection efficiency* η_{QE} [4]. The collection efficiency can be considered as the spectrally averaged quantum efficiency including reflection losses. The collection efficiency η_{QE} can be generally close to 100% for optimised cells. For non-concentrator silicon solar cells a value of 97% has been achieved [35]. In the TPV literature sometimes collection efficiencies in the system using the blackbody radiator are reported [28]. In these cases, it should be considered that η_{QE} can include effects such as the parasitic heat transfer to the cell (conduction/convection), the emissivity of the radiator of less than one and photons lost in the cavity (view factor less than one). Section 4.7.3 discusses further details of the difficulties to measure the cell performance with parasitic cavity effects.

$$\eta_{QE} = \frac{J_{ph}}{J_{max}} = \frac{J_{ph}}{e_0 \int_{\nu_g}^{\infty} n_{ph(\nu)} d\nu} = \frac{J_{ph}}{e_0 \int_{\nu_g}^{\infty} \frac{P_{PV(\nu)}}{h\nu \cdot A} d\nu} \tag{4.14}$$

4.2.4 Voltage Factor

The maximum voltage, which may be obtained from the PV cell, is the bandgap voltage V_g defined as the ratio of the PV cell bandgap energy $h\nu_g$ to the elementary charge e_0 (Eq. 4.16) [10]. Setting $J = 0$ in Eq. 4.2 gives the open circuits voltage V_{oc} as described in Eq. 4.15, where V_c is the thermal voltage. This equation shows that V_{oc} can be raised in two ways: by either increasing J_{ph} or decreasing J_s [27]. The photocurrent densities J_{ph} can be increased by optical concentration of radiation, where J_{ph} rises linearly with the concentration factor. This effect is beneficial for both, solar concentrator PV and TPV with high radiation densities compared to conventional solar PV. The second way is the reduction of J_s with

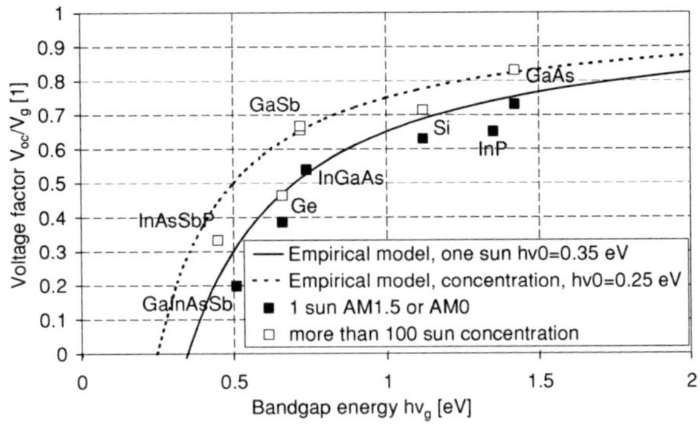

Fig. 4.3 Voltage factor vs. bandgap for different concentration levels, the figure was generated from different literature sources [27–32, 36]

some fundamental lower limits as discussed in Sect. 4.2.2 [26, 27]. In general, lower bandgap PV cells have higher dark saturation current densities J_s and hence lower open circuit voltages (compare Fig. 4.2) [26].

The ratio of open circuits voltage V_{oc} to the bandgap voltage V_g is defined as the *voltage factor* η_{OC} (Eq. 4.16) [4]. The voltage factor is usually the lowest and most critical efficiency of the PV cell related efficiencies (η_{OC}, η_{QE} and the fill factor η_{FF}) [9]. The voltage factor η_{OC} is a function of several parameters as discussed by Shockley and Queisser [10]. Figure 4.3 shows the voltage factor of different PV cells versus the bandgap energy. It can be seen that in particular low bandgap cells lack high voltage factors. This aspect is generally not beneficial since TPV cells typically have low bandgaps. It can be seen that GaSb, Ge and InGaAs PVcells are low bandgap cells with a high performance in terms of the voltage factor. Currently, there are no cells with a high voltage factor below a bandgap of about 0.7 eV. Figure 4.3 also presents a simple empirical model defined in Eq. 4.17 in this book. This model can describe the reduction of the voltage factor η_{OC} for a decrease in the bandgap energy and was fitted to the best performing cell for 1 sun operation and concentration factors larger than 100. The simplified model follows a more detailed model by Partain using the quasi fermi energy spacing from the nearest band edge, carrier concentrations and effective masses [27]. Figure 4.3 also shows that the voltage factor η_{OC} increases under high radiation densities. TPV cells typically operate under high illumination levels and hence this mechanism is beneficial. Comparing a silicon cell (1.1 eV) operating under non-concentrated solar radiation (AM1.5) with a GaSb cell (0.72 eV) operating under concentration (1473 K blackbody), it can be seen that similar voltage factors have been achieved due to the compensation of the two effects.

$$V_{oc} = \frac{kT_{cell}}{e_0} \ln\left(\frac{J_{ph}}{J_s} + 1\right) = V_c \ln\left(\frac{J_{ph}}{J_s} + 1\right) \approx V_c \ln\left(\frac{J_{ph}}{J_s}\right) \tag{4.15}$$

$$\eta_{OC} = \frac{V_{oc}}{V_g} = \frac{e_0 \cdot V_{oc}}{hv_g} \tag{4.16}$$

$$\eta_{\mathrm{OC,empirical}} = \frac{h\nu_g - h\nu_0}{h\nu_g} \tag{4.17}$$

Equation 4.18 approximately describes the drop of V_{oc} with *temperature* (negative temperature coefficient). The efficiency of the cell also drops with temperature because the voltage drop is dominant compared to the increase in the photocurrent. Other authors discuss in more detail the effect of temperature on V_{oc} [21, 37–39].

$$\frac{\mathrm{d}V_{\mathrm{oc}}}{\mathrm{d}T} \approx \frac{V_{\mathrm{oc}} - V_g}{T} \tag{4.18}$$

4.2.5 Fill Factor

The current-voltage curve of the PV cell under illumination has an exponential characteristic (diode curve in Fig. 4.1). The maximum electrical power density $J_m \cdot V_m$ is less than the product of $J_{\mathrm{ph}} \cdot V_{\mathrm{oc}}$, which is described by the *fill factor* η_{FF} Eq. 4.19.

$$\eta_{\mathrm{FF}} = \frac{J_m \cdot V_m}{J_{\mathrm{ph}} \cdot V_{\mathrm{oc}}} \tag{4.19}$$

In the absence of series resistance, the fill factor depends solely on V_{oc}/V_c and increases with this ratio [15]. Equation 4.20 defines the theoretical limit given by Shockley and Queisser [10]. Green proposed a simpler model (Eq. 4.21) [1]. Figure 4.4 shows that, for $V_{\mathrm{oc}}/V_c \geq 5$ the model by Green can describe the theoretical limit closely. A value of $V_{\mathrm{oc}}/V_c = 5$ corresponds to $V_{\mathrm{oc}} = 0.13$ V assuming a cell temperature of 25°C. Hence, the simpler model by Green can sufficiently describe cells currently of interest for TPV conversion ($h\nu_g > 0.4$ eV). Both equations show that the fill factor η_{FF} decreases for lower open circuit voltages (or lower PV cell bandgaps) and higher cell temperatures. Figure 4.4 illustrates that an increase of V_{oc} is accompanied by an improvement of the fill factor. It can be seen that high-performance cells show properties near the theoretical Shockley and Queisser model [27–32, 36].

$$\eta_{\mathrm{FF,Shockley}} = \frac{\left(\frac{V_m}{V_c}\right)^2}{\left(1 + \frac{V_m}{V_c} - e^{-\frac{V_m}{V_c}}\right)\frac{V_{\mathrm{oc}}}{V_c}} \tag{4.20}$$

$$\text{with: } V_c = \frac{kT_{\mathrm{cell}}}{e_0} \text{ and } \frac{V_{\mathrm{oc}}}{V_c} = \frac{V_m}{V_c} + \ln\left(1 + \frac{V_m}{V_c}\right)$$

$$\eta_{\mathrm{FF,Green}} = \frac{\frac{V_{\mathrm{oc}}}{V_c} - \ln\left(\frac{V_{\mathrm{oc}}}{V_c} + 0.72\right)}{\frac{V_{\mathrm{oc}}}{V_c} + 1} \tag{4.21}$$

Fig. 4.4 Fill factor vs. normalised open circuit voltage [10, 27, 28, 30–32, 36]

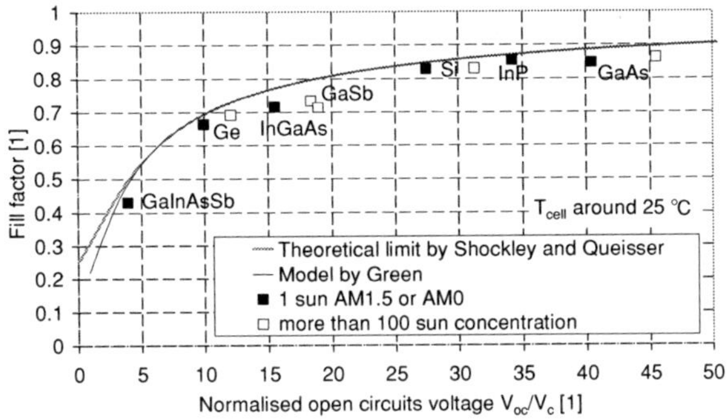

High-performance cells are designed with a low series and high parallel resistance values. Both, large series and small parallel resistance values, or their combined effects, lead mainly to a reduction of the fill factor [4, 13, 16, 27]. For small photocurrents, the fill factor depends solely on V_{oc}/V_c and increases slowly with it. At high photocurrents, or high radiation densities, the fill factor decreases due to series resistance losses. The maximum efficiency of a cell with a given series resistance occurs roughly where the product of current and series resistance equals the thermal voltage V_c [38]. Figure 4.4 shows that practical fill factor values of concentrator cells are similar compared to non-concentrator cells.

The fill factor generally decreases with increasing temperature. Martinelli and Stefancich discuss the impact of temperature in more detail [40].

4.2.6 Ideal PV Cell Related Efficiency

The previous three subsections discussed the collection efficiency η_{QE}, the voltage factor η_{OC} and the fill factor η_{FF} as the efficiencies related to the PV cell. In this subsection, the product of these three efficiencies is defined as the ideal PV cell related efficiency Eq. 4.22–4.25. For η_{QE}, or the mean quantum efficiency, there exists no clear dependency on the PV cell bandgap energy. Hence, for simplicity the quantum efficiency is assumed as unity, since values close to one can be achieved. For the voltage factor an empirical model with the bandgap as parameter has been presented (Sect. 4.2.4). For η_{FF}, the assumptions of the equation by Green limit the calculations to $h\nu_g > 0.4$ eV. Finally, it is possible to calculate η_{FF} depending on the bandgap energy, since V_{oc}, as a function of the bandgap is known from Eq. 4.23.

$$\eta_{QE} = 1 \tag{4.22}$$

$$\eta_{OC,empirical} = \frac{V_{oc}}{V_g} = \frac{e_0 V_{oc}}{h\nu_g} = \frac{h\nu_g - 0.25eV}{h\nu_g} \tag{4.23}$$

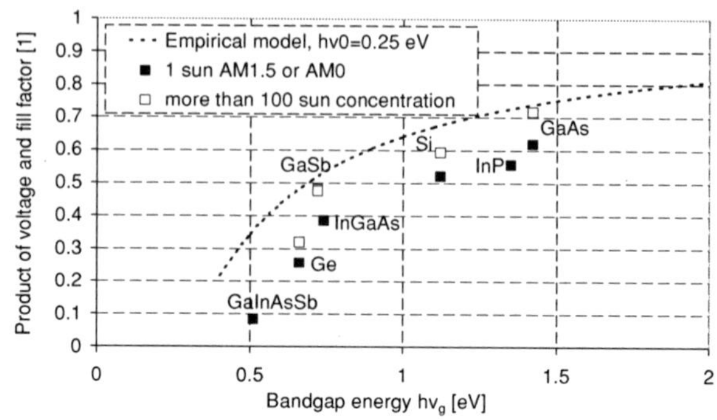

Fig. 4.5 Product of voltage and fill factor vs. bandgap energy [27, 28, 30–32, 36]

$$\eta_{\mathrm{FF,Green}} = \frac{v - \ln(v + 0.72)}{v + 1} \quad \text{with } v = \frac{V_{\mathrm{oc}}}{V_{\mathrm{c}}} = \frac{V_{\mathrm{oc}} e_{\mathrm{o}}}{kT_{\mathrm{cell}}} \tag{4.24}$$

$$\eta_{\mathrm{OC,FF,QE}} = \frac{v - \ln(v + 0.72)}{v + 1} \frac{hv_{\mathrm{g}} - 0.25eV}{hv_{\mathrm{g}}} \tag{4.25}$$

$$\text{with } v = \frac{hv_{\mathrm{g}} - 0.25eV}{kT_{\mathrm{cell}}}$$

Figure 4.5 shows the ideal PV cell related efficiency $\eta_{\mathrm{OC,FF,QE}}$. It can be seen that the cell related efficiency drops considerably towards smaller bandgap energies. This behaviour is caused by the fact that both, the voltage factor and the fill factor drop for lower bandgap cells.

Figure 4.5 can be compared to the emission peak of a blackbody at 0.46 eV (800°C) and 0.89 eV (1800°C). Hence, typical TPV operation requires PV cells with low bandgaps in this range. For a blackbody radiation distribution, it needs to be additionally considered that only 25% of the radiation is above the peak energy (e.g., 0.46 eV and 0.89 eV) [8]. Hence, even if the cell bandgap matches the emission peak of the blackbody about 75% of radiation cannot be converted! As a consequence typical blackbody radiation distributions and high-performance PV cells have a spectral mismatch. This is due to the fact that currently efficient PV cells are limited to higher bandgaps (roughly > 0.7 eV) and below this bandgap there is still a large share of photons that cannot be converted. Hence, it is very important to spectrally control radiation in a TPV cavity in order to achieve high system efficiencies. Section 6.2 describes spectral control theoretically in terms of the ultimate efficiency η_{UE}.

4.3 Fabrication Technologies and Epitaxial Growth

Several fabrication technologies to manufacture PV cells of the periodic groups IV (Si, Ge) and the combination of group III (Al, Ga, In) and group V (P, As, Sb) are known. The pn-junction of the PV cell can be formed by a diffusion technology,

Table 4.1 Overview of lattice matched substrate and epitaxial layer with low bandgaps

Substrate	Epitaxial layer	Bandgap (eV)	Comments	Literature
GaSb	InAsSb	0.29	Theoretical option	[26]
InAs	InAsSbP	0.35–0.5	Early research, low voltage	[3, 25, 26, 50–52]
InAs	InGaAsSb	0.3–0.7	Better performance on GaSb substrate	[51]
GaSb	InGaAsSb	0.5–0.6	High performance demonstrated	[3]
InP	$In_{0.53}Ga_{0.47}As$	0.74	Mature technology (e.g., fibre optics)	[3, 4]

epitaxial growth or a combination of these two technologies. The diffusion process is simpler and cheaper compared to the epitaxial growth process. Diffusion cells of interest for TPV conversion can be directly processed with their available substrates. These substrates are Si [27, 41], Ge [42], GaSb [26, 43, 44] and potentially InGaSb [26].

If a bulk crystal is not available, several epitaxial growth methods are available to form the pn-junction. These include metal organic vapour phase epitaxy Metal organic vapour phase epitaxy (MOCVD), also known as organometallic vapour phase epitaxy (MOVPE), liquid phase epitaxy (LPE) and to some extend also molecular beam epitaxy (MBE). As a substrate the group IV-semiconductors Ge and Si can be used. Currently, large 6″ wafers are available for bulk crystal Si and Ge. Also, the III-V semiconductor GaAs and InP are available as 6″ wafers, whereas materials such as GaSb and InAs are limited to smaller wafer sizes (2″) [45]. Besides the size and costs of wafers, the quality of the substrate has an impact on the cell performance. The bulk single crystal compounds InGaSb, InGaAs and InGaAsSb are difficult to grow from the melt using the traditional methods. There has been some effort towards the bulk crystal growth of these compounds [4, 26, 46–48].

High quality epitaxial layers require a careful selection of the substrate. Ideally, the atomic lattice constant of the grown layer matches the lattice constant of the substrate closely (lattice matched cell). Grown cell structures with a lattice mismatch to the substrate have typically a lower performance. Graded mismatched layers in between the cell and the substrate can overcome this problem to some extent. This was successfully demonstrated for InGaAs cells. Table 4.1 shows possible III-V semiconductor pairs of ternary and quaternary epitaxial layers lattice matched to the substrate. The table includes only low bandgap cells that are of interest for TPV conversion. For solar PV applications, such as space or concentrator systems, other lattice matched epitaxial grown cells are known. Examples are GaAs on Ge and GaInP on GaAs as single and mulijunction cells [43, 49]. Figure 4.6 visualises the correlations between the lattice constant and the bandgap. Si and Ge are indirect semiconductors and the other low bandgap III-V compounds of interest for TPV conversion are direct semiconductors.

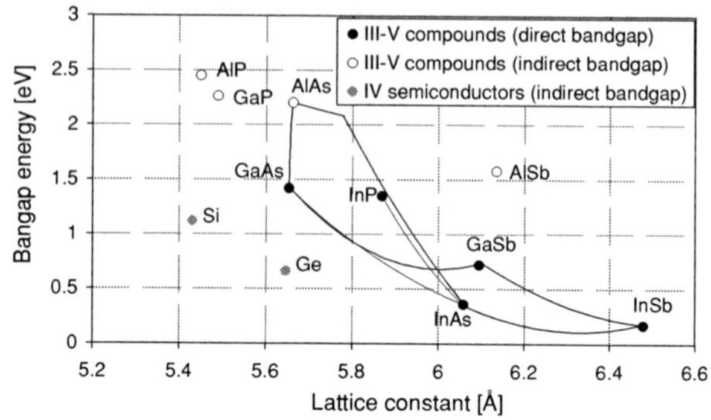

Fig. 4.6 Energy bandgap versus lattice constant of III-V semiconductors. The bandgap of silicon and germanium is also shown

4.4 PV Cell Design with Spectral Control

4.4.1 Front Surface Filters (FSFs)

Commonly PV cells are antireflection coated to couple radiation from a low refractive index media (air or vacuum) into the semiconductor with a high refractive index. This can be seen as a simple FSF [53]. For TPV, FSFs need to transmit in-band radiation and reflect mid and far infrared radiation. One advantage of FSFs is that they are cooled together with the PV cell, in contrast to filters placed in the TPV cavity, which may require additional cooling. All dielectric FSFs have been examined since the 1960 s for germanium cells [53]. All major filter types have been researched as FSF. These filters are FSS [54], TCO [55], all-dielectric [56], metal-dielectric [57] and tandem dielectric-TCO [54, 56].

4.4.2 Back Surface Reflectors (BSRs) and Buried Layer Reflectors

The technology of BSRs is partly available from space solar cells and silicon cells, where thin active layers and multiple passes of photons (light trapping) can increase the cell efficiency [2, 58]. For TPV, the principle of the BSR is that out-of-band radiation penetrates into the PV cell, passes through the device layer and substrate, is then reflected at the back surface mirror, returns via the same path and then leaves the PV cell to be returned to the radiator [1]. Hence, this concept requires low absorption of sub-bandgap radiation by the device layer and substrate. Low absorption due to free carriers can be achieved using low-doped substrates such as GaSb, InP and Si [1, 34, 59]. However, making an ohmic contact to a lightly doped substrate is generally difficult [1, 34, 60]. PV cells operating in TPV systems usually operate under high illumination intensities and this requires a particularly low series resistance to minimise ohmic losses. Besides a low resistance contact and low substrate absorption of the cell, other requirements for BSRs are high specular reflection, good adhesion, as well as good chemical and thermal stability [1, 61]. These requirements have lead to various PV cell designs. The reflector can be incorporated within (buried layer reflectors) or at the back surface the cell. Work on Ge, InGaAs and InGaAsSb cells has been reported, where in particular InGaAs cells achieved a very high out-of-band reflectivity [3, 51, 61, 62]. The potential of combining BSRs and FSFs has been pointed out by Coutts [1].

4.5 Group of IV Semiconductors

4.5.1 Silicon (Si)

Several TPV prototypes have been developed using silicon PV cells [63–66]. Silicon cells operating under high illumination are commercially available for solar concentrator PV systems [67]. Laser grooved buried contact silicon cells, widely used for non-concentrator solar conversion, have been reported to be also suitable for higher concentrations [2]. Silicon has an indirect bandgap of $h\nu_g = 1.12$ eV at 300 K (responds up to 1.11 μm) [16]. For TPV conversion, this bandgap is usually regarded as high. Typically silicon cells have been used with an ytterbia (Yb_2O_3) radiator (e.g., Welsbach mantle type), which has a suitable selective emission for the silicon bandgap [42]. References about back surface reflector (BSR) work in conjunction with silicon cells have been summarised by Charache et al. [61]. After this summary, Hampe et al. at ISFH reported about a metal insulation semiconductor (MIS) silicon PV cell using a BSR [34, 68]. For solar PV conversion, a small record silicon cell has approximately the following performance: $\eta_{PV} = 25\%$ with $\eta_{OC} = 63\%$, $\eta_{QE} = 97\%$, $\eta_{FF} = 83\%$ and $\eta_{UE} = 49\%$. This calculation assumed the global AM1.5 spectrum with a total intensity of 0.1 W/cm^2 (1 sun illumination) [35][1]. A chapter by Sinton and Blakers gives a good overview of silicon concentrator solar cells with developments such as back contact, buried contact and silver cells for high solar radiation densities. TPV converters could also utilise such silicon cells, but to the authors' knowledge these cells have been not considered [27, 41].

[1] The values were calculated within this work. The value η_{OC} was calculated as $e_0 \cdot V_{OC}/h\nu_g$ = 0.706/1.12 = 0.630. Using AM1.5 data from [19] the maximum photocurrent of J_{max} = 43.6 mA/cm^2 was computed (also given in [27]). This results in η_{QE} = 42.2/43.6=0.968. The ultimate efficiency was computed as η_{UE} = 0.488 for the global AM1.5 spectrum. This value is slightly higher than the blackbody value of 0.44 [10].

4.5.2 Germanium (Ge)

In the early stages of TPV research, germanium PV cells were utilised. However, these cells suffered from poor cell performance [63]. Recently, more advanced germanium PV cells have been developed as a bottom cell for multijunction solar converters. The Ge substrate for the TPV cells is of great interest because of its low cost. The costs of germanium substrates are reported to be lower than those of GaSb by a factor of 6–7 [29, 32, 42, 45, 69, 70]. Germanium has an indirect bandgap of $h\nu_g = 0.66$ eV at 300 K and responds up to 1.88 μm [16]. Germanium cells spectrally match the selective emission of erbia radiators [42]. A typical germanium cell has approximately the following performance: $\eta_{PV} = 6.7\%$ with $\eta_{OC} = 37.1\%$, $\eta_{QE} = 75.2\%$, $\eta_{FF} = 59.6\%$ and $\eta_{UE} = 40.2\%$. The calculation assumed the global AM1.5 spectrum and an intensity of 0.1 W/cm^2 [69][2]. An improved cell with $\eta_{PV} = 8.4\%$ efficiency (1 sun AM1.5G) has been later presented by the same laboratory [32]. These partial efficiencies show that in particular high voltage factors are one major challenge for germanium cells [26]. On the other hand, Nagashima simulated a back contact type cell with voltage factors well over 50% for high radiation densities. For germanium cells, BSRs are generally feasible and different authors discussed these developments [3, 29, 32, 62].

4.5.3 Silicon-Germanium (SiGe)

SiGe structures are used for high-frequency semiconductor devices. SiGe alloys allow changing the bandgap between germanium (0.66 eV) and silison (1.12 eV) by varying the composition. Palfinger considered such cells for TPV conversion [71]. For more information, the reader is referred to Bitnar who published an overview article about Si, Ge and Si/Ge-cells [42].

4.6 Group of III-V Semiconductors

PV cells based on III-V semiconductors have currently the highest conversion efficiencies and they are utilised for solar power in space [31, 49, 72]. Terrestrial solar concentrator systems offer a large potential market. The high cost of III-V semiconductors is still a major hindrance to their widespread use. For cost reduction polycrystalline cells [73, 74], thin film cells [75–77] and growth on

[2] The value η_{OC} was calculated through $e_0 \cdot V_{OC}/h\nu_g = 0.245/0.66 = 0.371$ using AM1.5 data from [19]. A maximum photocurrent of $J_{max} = 60.9$ mA/cm^2 was computed (also given in [27]), which gives $\eta_{QE} = 45.8/60.9 = 0.752$. The ultimate efficiency was computed as $\eta_{UE} = 0.402$ for the global AM1.5 spectrum.

cheap substrates are of interest [26, 78]. For TPV conversion, the major low bandgap materials currently of interest are gallium antimonide (GaSb), indium gallium arsenide (InGaAs) and indium gallium arsenide antimonide (InGaAsSb). More recently InAsSbP layers on InAs substrate have been examined. The bandgap of InAsSbP cells between 0.35 and 0.5 eV could be even lower, but the open circuit voltages of these devices were also very low. Bett [43, 44], Wang [51], Andreev [3], Wanlass [79] and Mauk [25, 26] gave overviews of low bandgap III-V PV cells. In particular the book chapter by Mauk gives a comprehensive overview [25].

4.6.1 Gallium Antimonide (GaSb)

At present, GaSb PV cells are often regarded as the most suitable choice for TPV generators [43, 44]. This material has a direct bandgap of 0.72 eV (1.72 μm) at 300 K [16]. Several research groups and companies have manufactured GaSb cells using diffusion or epitaxial technologies (MOCVD and LPE) [3, 26, 44, 51]. Historically, GaSb cells have been developed as a bottom cell for mechanically stacked tandem solar PV converters (GaAs/GaSb). This tandem cell has demonstrated a high performance [35, 80]. In the early 1990's two companies were founded (Fraas at JX-Crystals Inc. and Horne at EDTEK Inc.). Both companies started to produce Boeing's Zn-diffused GaSb cell technology under license [63]. The zinc-diffusion process is well understood and other groups (e.g. Ioffe Russia, ISE Germany) demonstrated also GaSb cells with high performance [44, 81]. Spectral control approaches for GaSb cells have included (also in combination) selective radiators (erbia, matched cobalt radiator), thick SiO_2 windows, filters and PV cell front surface filters (FSF). The difficulties in incorporating BSRs into GaSb cells are discussed by Charache et al. [61]. The performance of a zinc-diffusion GaSb cell under 1473 K blackbody radiation can be calculated as follows[3]: $\eta_{PV} = 3\%$ with $\eta_{OC} = 65.6\%$, $\eta_{QE} = 47\%$, $\eta_{FF} = 73.3\%$ and $\eta_{UE} = 13.6\%$. The value for the cell efficiency η_{PV} can be also approximately confirmed by the measured electrical power density and the total blackbody radiation $\eta_{PV} = P_{el,meas}/\sigma T_s^4 = 0.82/26.7 \cdot 100\% = 3\%$ [28]. This efficiency value shows the vital importance of spectral control for TPV. Assuming all radiation below the bandgap energy (0.72 eV) can be recovered (not absorbed in the cell), the ultimate efficiency, and hence the cell performance, improves significantly: $\eta_{UE} = 80.9\%$ and $\eta_{PV} = 18\%$. It can be seen that high fill factors η_{FF} and high voltage factors η_{OC} can be achieved for GaSb cells. The low collection efficiency of $\eta_{QE} = 47\%$ can be considered to be the result of the cavity design (view factor, radiation angle), where incident photons

[3] The value η_{OC} was calculated through $e_0 \cdot V_{OC}/h\nu_g = 0.472/0.72 = 0.656$. Using a 1473 K blackbody spectrum the maximum photocurrent of $J_{max} = 5.06$ A/cm^2 was computed, which gives $\eta_{QE} = 2.36/5.06 = 0.47$. The ultimate efficiency was computed as $\eta_{UE} = 0.136$ for the 1473 K blackbody spectrum and the bandgap of 0.72 eV.

under low angles are totally reflected at the cell surface and other photons may have been lost in between the radiator and the PV cells. From the quantum efficiency versus wavelength curve [3, 28] a value around $\eta_{QE} = 80\%$ can be expected, which yields an even higher cell efficiency: $\eta_{PV} = 31\%$. In summary, it can be said that the cell related efficiency (product of η_{OC}, η_{QE} and η_{FF}) is currently usually smaller for GaSb operating in a TPV system (39%) compared to high-performance silicon cells converting solar radiation (51%). However, the total cell efficiency η_{PV} can be higher for TPV due to a higher ultimate efficiency achieved by using suitable spectral and angular control methods. At the moment there is no large market for GaSb cells and hence cell costs are high. Developments in the solar concentrator market using GaSb bottom cells could lead to improved economics.

4.6.2 Indium Gallium Arsenide (InGaAs)

The bandgap of $In_xGa_{1-x}As$, also called GaInAs, can be varied in a wide range from 1.42 to 0.36 eV (0.87–3.44 μm) by varying x from 0 to 1 [4, 79, 82]. Figure 4.6 shows the bandgap range as a line between GaAs and InAs versus lattice constant. Cell structures in *lattice matched* $In_{0.53}Ga_{0.47}As$ on InP substrate (bandgap 0.74 eV, 1.68 μm) have been manufactured mainly by epitaxial methods (MOCVD and LPE) [3], but also a few by diffusion methods [83]. Fabrication of cells based on InGaAs is a mature technology and applications include infrared photo detectors (e.g., fibre optic communication), laser power converters and bottom cells in solar tandem converters [3, 4]. Wilt et al. reported the performance of a 0.74 eV InGaAs cell under the AM0 spectrum [30]. The efficiency values can be calculated as follows[4]: $\eta_{PV} = 11.9\%$ with $\eta_{OC} = 53.9\%$, $\eta_{QE} = 75.3\%$, $\eta_{FF} = 71.5\%$ and $\eta_{UE} = 40.9\%$. Ginige et al. reports about the adaptation from laboratory to industrial batch manufacturing [84].

For TPV conversion lower bandgap *lattice mismatched InGaAs* cells are of interest. These cells can be produced by increasing the indium and decreasing the gallium content. Different lattice mismatched $In_xGa_{1-x}As$ cells on an InP substrate with bandgaps from 0.6 eV (2.07 μm) to 0.55 eV (2.25 μm) have been grown by MOCVD [3, 85]. A high performance has been demonstrated for the 0.6 eV $In_{0.68}Ga_{0.32}As$ cell ($\eta_{OC} = 59\%$ and $\eta_{FF} = 73.4\%$) under concentration [86]. Cells with higher lattice mismatch (lower bandgaps) on InP substrates appear to be difficult to manufacture with high performance. TPV devices fabricated in InGaAs expitaxial structures on InP substrates is represented by work at NREL (Cleveland), NASA-Glenn Research Center (Cleveland), Bechtel Bettis (Pittsburgh), Ohio State University and others [25].

---

[4] The value η_{OC} was calculated through $e_0 \cdot V_{OC}/h\nu_g = 0.3988/0.74 = 0.539$. Using AM0 spectrum data from (1353 W/m^2) a maximum photocurrent of J_{max} = 74.8 mA/cm^2 was computed, which gives η_{QE} = 56.35/74.8=0.753. The ultimate efficiency was computed as η_{UE} = 0.409 for the AM0 spectrum.

Monolithic Interconnected Modules (MIMs) have been designed using lattice matched and mismatched $In_xGa_{1-x}As$ cells since 1994 [30, 87, 88]. These MIMs are typified by an InP semi-insulating substrate [30]. This substrate provides two key advantages: electrical isolation for the component cells and near perfect optical transparency for sub-bandgap photons. The former allows series connection to increase the voltage and the latter allows spectral control using a BSR [30]. Wilt et al. discusses other advantages of the MIM structure [30]. Tandem cells using the MIM structure have also been developed. They consist of a 0.72 eV lattice matched and a lower bandgap mismatched cell [241]. MIMs are generally suitable for broadband radiators due to the incorporation of back surface or buried reflectors with a high reflectivity [3, 88–90]. Although, for longer wavelengths (e.g., at 10 μm) a significant reduction of the reflectivity is obtained. Hence, an additional FSF, as proposed by Coutts, can improve the efficiency further. Wernsman reported about a combined dielectric-TCO filter glued with an optical epoxy onto the MIM [1, 91].

4.6.3 Indium Gallium Arsenide Antimonide (InGaAsSb)

Quaternary III–V alloys afford two degrees of freedom in tuning both the bandgap and lattice constant. This is often necessary in order to produce an epitaxial layer with the desired bandgap *and* which is closely lattice matched with its substrate [26]. InGaAsSb (also called GaInAsSb) cells can be lattice matched on a GaSb and InAs substrates with theoretical bandgaps from 0.29 to 0.72 eV. A higher thermodynamic stability of the alloy is achieved for epilayers lattice matched to GaSb. Hence, GaSb substrates have been utilised [51]. The miscibility gap limits the lower bandgap to about 0.5 eV [26]. InGaAsSb cells on GaSb substrates have been fabricated with bandgaps from approximately 0.5 to (2.48 μm) to 0.6 eV (2.07 μm) using diffusion and epitaxial growth processes including LPE, MOCVD and MBE [3, 26]. InGaAsSb cells (0.52 eV) with voltage factors of $\eta_{OC} = 57.7\%$ and fill factors of $\eta_{FF} = 70\%$ have been reported [92]. The challenges in designing InGaAsSb MIM structures are discussed elsewhere [26, 87]. Cells with buried reflectors have also been examined [59, 93]. TPV devices fabricated in InGaAsSb expitaxial structures on GaSb substrates is represented by work at MIT Lincoln Labs (Boston), Sarnoff Labs (Princeton), Sandia National Laboratories (New Mexico), KAPL/Lockheed-Martin (Schenectady), the Ioffe Institute (St. Petersburg), the Fraunhofer ISE (Freiburg) and other groups [25].

4.7 Other Materials and Aspects

4.7.1 Tandem Cells

One major loss in the efficiency of a single junction cell arises from two mechanisms described by the ultimate efficiency. The first mechanism is that photons with energies hv smaller than the bandgap energy $h\nu_g$ are not converted (free carrier heating). TPV conversion allows suppression of free carrier heating by some means of spectral control. The second mechanism is due to photons with energies $h\nu > h\nu_g$. These photons contribute only to the energy output $h\nu_g$ (hot carrier heating). Theoretically, this loss mechanism could be also minimised by spectral control methods, but this would result in low, often not acceptable, power densities. It can be argued that spectral control means are more effective than tandem cells to achieve high conversion efficiencies, if currently available efficient cells ($h\nu_g > 0.5$ eV) and basic ultimate efficiency modelling (see Sect. 6.2) are considered. Hence, tandem cells can be seen as a further, but not the first path to improve conversion efficiency and maintain a high power density in TPV conversion by reducing hot carrier heating. The principle of tandem cells, or multiple-junction or cascaded cells, is to subdivide the broad radiation spectrum into two or more spectral bands and to convert each band with a cell with a matched bandgap. There are also some other methods to split the spectrum. They include spectrally selective mirrors, holographic filtering and refractive filtering. Imenes et al. gave a review of these split spectrum methods (see also Sect. 3.8) [94]. For TPV conversion, these methods are currently not considered.

The idea of tandem cells seems to have been first suggested in 1955 and it was already realised that spectral filtering could be simply achieved, merely by stacking cells on top of one another, with the largest bandgap cell uppermost [95, 96]. The top cell absorbs all the photons at and above its bandgap energy and transmits the less energetic photons to the cells below [72]. For stacked cells, there are two approaches, namely mechanically stacked single junction cells and epitaxially grown monolithically integrated cells. The monolithic tandem cell has the advantage that only one substrate is required. On the other hand the growth is more complex. Tandem cells differ in the number of electrical terminals (2, 3 or 4 terminal). Typically it is intended to operate the cell with two terminals to reduce the complexity.

Various types of high-performance tandem *solar* PV cells utilise Ge, GaSb and InGaAs as a bottom cell for space and concentrator applications [31]. These technological developments in solar tandem cells are very beneficial towards low bandgap cells with a high quality for TPV conversion. In particular the solar concentrator developments are of interest, since these cells are already designed for high currents. For example, mechanically stacked GaSb cells for solar concentrator systems are applicable for TPV systems.

At the current status of TPV conversion, the developments aim for epitaxially grown monolithically integrated cells with two junctions and two terminals. Hence, the following text is restricted to this type. The series connected two-terminal structure has the disadvantage that the photocurrent of both cells has to match and this limits the range of bandgap energies for the top and bottom cell. Furthermore, the performance of the two-terminal structure is affected by changes of the cell environment. Changes of the cell temperature and the radiation spectrum can cause a mismatch of the current and ultimately lead to a lower performance of the cell [72, 95, 97]. It can be argued that matching of the radiation spectrum and the cell bandgaps can be engineered in a TPV system carefully, compared to the variable solar spectrum. An advantage of the series connected structure is that the voltage increases and the current decreases with the number of stacked cells. For low voltage cells and high illumination levels, as for TPV conversion, this leads to lower series resistance losses. For TPV conversion, different monolithically integrated tandem cells with low bandgaps have been epitaxially grown. The III–V semiconductors include GaSb/InGaAsSb (0.72/0.56 eV) [3, 26, 98], InGaAs/InGaAs (0.74/0.63 eV) [79, 87, 99] and InGaAsP/InGaAs (0.72/0.60 eV) [100].

4.7.2 Alternative Semiconductors, Cell Designs and Concepts

Some research on *InGaSb* cells grown on GaSb substrate has been reported. There is a larger mismatch of the atomic lattice of the epitaxial grown InGaSb layer on the GaSb substrates. Hence, the research focussed mostly on fundamental growth characterisations [1, 4, 26, 50].

Since 1990, quantum well cells (QWCs) have been examined. A QWC is a photovoltaic device with a multiple quantum well system inserted in the intrinsic region of a p–i–n cell. QWC has a tuneable bandgap and this allows growth of cells made from material with a much wider range of compositions for which no lattice matched substrates exist. Also, the efficiency of QWC should be less sensitive to an increased cell temperature compared to bulk cells. The developments aim for InGaAsP and InGaAs QWC on InP substrate [101–103].

For terrestrial solar concentrator systems, *vertical junction*, or silver, cells using silicon have been developed [41]. Individual cells can be connected in series to build up high voltages and obtain low currents. In this way series resistance losses can be minimised at high radiation densities. Sater discusses the utilisation of vertical junction germanium cells using a back surface reflector [104].

For solar PV, *thin film* cells are considered one major option to reduce PV module prices in future (e.g., CdTe, $CuInGaSe_2$, Si) [31]. At the current stage thin film cells play a minor role in TPV conversion. This situation could change in long-term, as for solar PV [1, 71, 76, 77].

For most TPV systems, the PV cell temperature is maintained below 100°C. Luque proposed to *operate the solar cell hot* (e.g., up to 230°C) and to utilise the wasted heat of the cell in an additional heat engine, such as a thermoelectric generator [105]. Hot cell operation has been already considered for TPV cells operated in space. One major difficulty for systems used in space is the removal of heat from the PV cells. Low temperature operation would require large cooling fins, since heat can only be removed by radiation in space. Hence, TPV systems using PV cells at temperatures up to 230°C have been examined [106–108]. The major efficiency limitation originates from the decrease of the fill factor and voltage factor with increasing temperature. One advantage is that available PV cells with higher bandgaps can be utilised, since the bandgap energy decreases to a suitable lower value at high operation temperature. TPV high-temperature cell research may also gain from solar PV developments for near-sun missions.

Different approaches currently under investigation for solar PV conversion may also impact on TPV conversion in future, if these approaches can be fully demonstrated. Examples are hot electron cells [19, 45, 105], luminescent up and down converters [109] and solar collection by antenna-rectifier areas [110, 111]. The latter, antenna-rectifiers have demonstrated highly efficient conversion of microwave radiation into electricity. For example, systems capable of converting more than 80% of monochromatic microwave radiation into electricity have been demonstrated. For conversion of solar radiation one major challenge is the requirement of small physical structures. Infrared (IR) radiation would allow larger structures and the radiator spectrum may be also tailored narrowly. Hence, antenna-rectifier areas may be rather suited for the direct conversion of heat into electricity, but to the authors' knowledge there is no research in this field.

4.7.3 Efficiency Measurement of PV Cells

For solar flat-plate PV, procedures have been established and standardised for the determination of the cell efficiency. The reference conditions include the temperature, total radiation density, spectral distribution and the definition of the area. For solar concentrator PV, the standardisation is more difficult and very few standards exist or are being developed. For TPV conversion, there are no standard conditions defined to report the PV cell efficiency at the moment [45]. The similarities of solar concentrator PV and TPV have been pointed out and may help to identify suitable TPV procedures [112]. The efficiency depends strongly on the radiation spectrum and TPV cells have been characterised using various spectra. They include solar PV conditions (AM0, AM1.5) and TPV radiators (e.g., silicon carbide or tungsten). Not only the spectral radiation distribution but also the spatial and angular radiation distributions have an impact on the cell efficiency. The spatial radiation distribution refers to the uniformity of illumination. The angular radiation distribution refers to the incidence angle of the radiation onto the cell. An ideal blackbody radiator emits radiation with a lambertion distribution and this is fundamentally different from collimated solar radiation. Hence, the cell efficiency depends on the angular distribution of the radiation. In a TPV system, the spatial, angular and spectral radiation distributions are linked to the cavity design.

This brief discussion points out that the efficiency measurement of TPV cells is a challenging task. Hence, reported cell efficiencies need to be assessed carefully because the values depend strongly on the conditions. Efforts to establish characterisation procedures are very valuable for TPV progress. An ultimate goal would be to establish standard characterisation conditions [45, 112–118].

4.7.4 PV Cell Cooling

Generally, PV cell cooling technologies are available from concentrator solar PV and power electronic systems. For the cell cooling, a suitable thermo-mechanical design of the cell-to-sink interconnection needs to be identified. The PV cell, electrical insulation layer and the thermally conductive substrate (e.g., copper or aluminium) differ in terms of their thermal expansion coefficient and this causes thermo-mechanical stress. Common electrical insulation layers that fulfil the requirements are epoxy layers and ceramic substrates, such as Al_2O_3, AlN or BeO. For GaSb cells, AlN and BeO have been used [81, 119]. A chapter by Martinelli and Stefancich discusses the cell-to-sink interconnection aspects in more detail [40].

Heat exchangers, connected to the substrate, can use a gas (typically air) or a liquid (typically water) to absorb the heat. In general, passive (e.g., free air convection, thermo-siphon [120], heat pipes) and active systems (e.g., forced air [121] or water) are feasible. Typically PV cells are cooled by a forced convection water cycle using a pump. Complete systems also require an air heat exchanger for heat removal to the environment. In order to increase the heat transfer rate a liquid–gas phase change (boiling) in the heat carrier can be utilised. Carlson and Fraas describe a system using a boiling coolant for the cell and an air-cooled condenser to remove the heat to the ambient [122]. Cell cooling by the utilisation of the Joule Thomson effect of expanding liquefied fuel may be also utilised.

4.7.5 Auxiliary Electrical Components

TPV systems also require an electrical maximum power tracker in order to operate the PV cell array at its maximum power point (compare Fig. 4.1). Different trackers are available from solar PV applications. The operation conditions are usually less critical for steady illuminated cells in TPV systems compared to the fluctuating illumination conditions experienced for solar conversion.

Currently TPV systems are usually designed for constant electrical power output. For some application this operation mode may be suitable (e.g., industrial waste heat recovery system) and for others not (e.g., portable power supply with varying load). *Electrical energy storage* can overcome this limitation. For load levelling, secondary batteries have been considered as a supplement [121]. In principle, other options are also feasible, such as (super) capacitors.

4.8 Summary

This chapter presented some fundamental theory of PV cells. For TPV conversion, it is important to understand the impact of the cell bandgap energy on the conversion efficiency. For efficiency modelling, one may distinguish two major approaches. The first approach utilises a model for the dark saturation current density versus bandgap energy. The second approach, presented in this chapter in detail, is based on partial cell efficiencies. These efficiencies are the voltage factor, the fill factor, the collection efficiency and the ultimate efficiency. The advantage of partial cell efficiencies is the possibility to distinguish between the cell performance and the quality of the spectral control. It was shown that the cell related efficiencies (product of fill factor and voltage factor) drop significantly towards lower bandgap energies. On the other hand, the optimum ultimate efficiency typically requires low bandgap cells. Hence, for an efficient design, there is usually a trade-off between cell related efficiencies and the ultimate efficiency. The assessment of partial efficiencies allows the identification of critical efficiencies and the optimisation of the overall efficiency.

Of the group IV semiconductors, silicon cells are well understood and have demonstrated high performance. On the other hand, silicon cells have a relatively high bandgap for TPV conversion. For an efficient system, silicon cells require high radiator temperatures or a spectral control concept with a very high performance. At the current stage, suitable spectral control concepts have not been demonstrated. Germanium cells can be considered to compete with GaSb cells because of their similar bandgap energy. The performance of GaSb cells is higher, but recently improvements in Ge cells have been reported. Ge cells would have the advantages of lower substrate costs and they could include a back surface reflector for spectral control.

The major low bandgap III–V semiconductors of interest are GaSb (0.7 eV), InGaAs (0.6–0.7 eV) and InGaAsSb (0.5–0.6 eV), where the utilised bangap energies are given in brackets. Zn-diffused GaSb cells are well understood and large-scale production of such cells is feasible. More sophisticated epitaxially grown cells (InGaAs and InGaAsSb) also demonstrated high performances. These cells offer more flexibility to address specific TPV aspects compared to diffusion cells. For example, Monolithic Interconnected Modules (MIMs) allow for designs with increased cell voltages and additional spectral control means (e.g., buried reflectors). In addition, epitaxially grown III–V multijunction cells offer a promising path towards lower currents (or higher voltages) and improved efficiencies.

References

1. Coutts TJ (1999) A review of progress in thermophotovoltaic generation of electricity. Renew Sustain Energy Rev 3(2–3):77–184
2. Coutts TJ (2001) Chapter 11: Thermophotovoltaic generation of electricity. In: Archer MD, Hill R (eds) Clean electricity from photovoltaics, vol 1. Series on photoconversion of solar energy, Imperial College Press, London
3. Andreev VM (2003) An overview of TPV cell technologies. Proceeding of the 5th conference on thermophotovoltaic generation of electricity, Rome, Italy, 16–19 Sept 2002. American Institute of Physics, pp 289–304
4. Bhat IB, Borrego JM, Gutmann RJ, Ostrogorsky AG (1996) TPV energy conversion: a review of material and cell related issues. Proceeding of the 31st intersociety energy conversion engineering conference. IEEE, pp 968–973

5. Iles PA (1990) Non-solar photovoltaic cells. Proceeding of the 21st IEEE Photovoltaic Specialists Conference, IEEE, pp 420–425
6. Woolf LD (1986) Optimum efficiency of single and multiple bandgap cells In: thermophotovoltaic energy conversion. Sol Cells 19(1):19–38
7. Woolf LD (1985) Optimum efficiency of single and multiple band gap cells in TPV energy conversion. Proceeding of the 18th IEEE photovoltaic specialists conference. IEEE, pp 731–1732
8. Chubb D (2007) Fundamentals of thermophotovoltaic energy conversion. Elsevier Science, Amsterdam
9. Baldasaro PF, Brown EJ, Depoy DM, Campbell BC, Parrington JR (1995) Experimental assessment of low temperature voltaic energy conversion. Proceeding of the 1st NREL conference on thermophotovoltaic generation of electricity, Copper Mountain, Colorado, 24–28. July 1994. American Institute of Physics, pp 29–43
10. Shockley W, Queisser HJ (1961) Detailed balance limit of efficiency of p-n junction solar cells. Appl Phys 32:510–519
11. Yeargan JR, Cook RG, Sexton FW (1976) Thermophotovoltaic systems for electrical energy conversion. Proceeding of the 12th IEEE photovoltaic specialists conference, IEEE, pp 807–813
12. Charache GW, Egley JL, Depoy DM, Danielson LR, Freeman MJ, Dziendziel RJ, Moynihan JF, Baldasaro PF, Campbell BC, Wang CA, Choi HK, Turner GW, Wojtczuk SJ, Colter P, Sharps P, Timmons M, Fahey RE, Zhang K (1998) Infrared materials for thermophotovoltaic applications. J Electron Mater 27:1038–1042
13. Baldasaro PF, Raynolds JE, Charache GW, DePoy DM, Ballinger CT, Donovan T, Borrego JM (2001) Thermodynamic analysis of thermophotovoltaic efficiency and power density tradeoffs. Appl Phys 89:3319–3327
14. Murray S, Aiken D, Stan M, Murray C, Newman F, Hills J, Siergiej RR, Wernsman B (2002) Effect of metal coverage on the performance of 0.6 eV InGaAs monolithic interconnected modules. Proceeding of the 5th conference on thermophotovoltaic generation of electricity, Rome, Italy, 16–19 Sept 2002. American Institute of Physics, pp 424–433
15. Luque A (1990) The requirements of high efficiency solar cells. In: Luque A, Araujo GL (eds) Physical limitations to photovoltaic energy conversion. Taylor & Francis, London, pp 1–42
16. Sze SM (1981) Physics of semiconductor devices, 2nd edn. Wiley, New York
17. Coutts TJ, Ward JS (1999) Thermophotovoltaic and photovoltaic conversion at high-flux densities. IEEE Trans Electron Devices 46(10):2145–2153
18. Fraas L, Samaras J, Han-Xiang Huang, Seal M, West E (1999) Development status on a TPV cylinder for combined heat and electric power for the home. Proceeding of the 4th NREL Conference on Thermophotovoltaic Generation of Electricity, Denver, Colorado, 11–14. Oct. 1998. American Institute of Physics, pp 371–383
19. Würfel P (1995) Physik der Solarzellen (in German), 2nd edn. Spektrum Akademischer Verlag, Heidelberg
20. Henry CH (1980) Limiting efficiency of ideal single and multiple energy gap terrestrial cells. J Appl Phys 51:4494–4500
21. Létay G (2003) Modellierung von III-V Solarzellen (in German). Doctoral thesis, University of Konstanz
22. Charache GW, Baldasaro PF, Danielson LR, Depoy DM, Freeman MJ, Wang CA, Choi HK, Garbuzov DZ, Martinelli RU, Khalfin V, Saroop S, Borrego JM, Gutmann RJ (1999) InGaAsSb thermophotovoltaic diode: physics evaluation. J Appl Phys 85:2247–2252
23. Catalano A (1996) Thermophotovoltaics: a new paradigm for power generation? Renewable Energy 8:495–499
24. Wanlass MW, Emery KA, Gessert TA, Horner GS, Osterwald CR, Coutts TJ (1989) Practical considerations in tandem cell modeling. Sol Cells 27:191–204

25. Mauk MG (2006) Survey of thermophotovoltaic (TPV) devices. In: Krier A (ed) Mid-infrared semiconductor optoelectronics. Springer, London, pp 673–738
26. Mauk MG, Andreev VM (2003) GaSb-related materials for TPV cells. Semicond Sci Technol 18:191–201
27. Partain LD (1995) Solar cells and their applications. Wiley Interscience, New York
28. Sale Items (2010) JX-Crystals Inc., US [Online] Available at: http://www.jxcrystals.com/4salenew.htm. Accessed 28 April 2010
29. Andreev VM, Khvostikov VP, Khvostikova OV, Oliva, EV, Rumyantsev VD, Shvarts MZ, Tabarov TS (2003) Low band gap Ge and InAsSbP/InAs-based TPV cells. Proceeding of the 5th conference on thermophotovoltaic generation of electricity, Rome, Italy, 16–19 Sept 2002. American Institute of Physics, pp 383–391
30. Wilt DM, Fatemi NS, Jenkins PP, Weizer VG, Hoffman, RW, Jain RK, Murray CS, Riley DR (1997) Electrical and optical performance characteristics of 0.74 eV p_n InGaAs monolithic interconnected modules. Proceeding of the 3rd NREL Conference on thermophotovoltaic generation of electricity, Denver, Colorado, 18–21 May 1997. American Institute of Physics, pp 237–247
31. Green MA, Emery K, Hishikawa Y, Warta W (2009) Solar cell efficiency tables (version 33) short communication. Prog Photovoltaics Res Appl 17:85–94
32. van der Heide J, Posthuma NE, Flamand G, Poortmans J (2007) Development of low-cost thermophotovoltaic cells using germanium substrates. Proceeding of the 7th world Conference on the thermophotovoltaic generation of electricity, Madrid, 25–27 Sept 2006. American Institute of Physics, pp 129–138
33. Sellers I (2000) Quantum well cells for applications in thermophotovoltaics. PhD Thesis, University of London, Imperial College of Science, Technology and Medicine
34. Hampe C (2002) Untersuchung influenzierter und diffundierter pn-Übergänge von Terrestrik- und Thermophotovoltaik-Siliciumsolarzellen (in German), Doctoral thesis, Institut für Solarenergieforschung GmbH Hameln/Emmerthal (ISFH)
35. Green MA, Emery K, King DL, Igari S, Warta W (2003) Solar cell efficiency tables (version 22). Prog Photovoltaics Res Appl 11:347–352
36. Welser E, Dimroth F, Ohm A, Guter W, Siefer G, Philipps S, Schöne J, Polychroniadis EK, Konidaris S, Bett AW (2007) Lattice-matched GaInAsSb on GaSb for TPV cells. Proceeding of the 7th world conference on the thermophotovoltaic generation of electricity, Madrid, 25–27 Sept 2006. American Institute of Physics, pp 107–114
37. Zenker M (2001) Thermophotovoltaische Konversion von Verbrennungswaerme (in German). Doctoral thesis, Albert-Ludwigs-Universität Freiburg im Breisgau
38. Luque A (2001) Concentrator cells and systems. In: Archer MD, Hill R (eds) Clean electricity from photovoltaics, Chap 12, vol 1. Series on photoconversion of solar energy, Imperial College Press, London
39. Fraas LM, Avery JE, Gruenbaum PE, Sundaram S, Emery K, Matson R (1991) Fundamental characterization studies of GaSb solar cells. Proceeding of the 22nd IEEE photovoltaic specialists conference, pp 80–84
40. Martinelli G, Stefancich M (2007) Solar cell cooling. In: Luque A, Andreev V (eds) Concentrator photovoltaics, Chap 7. Springer, Berlin, pp 133–149
41. Blakers A (2007) Silicon concentrator solar cells. In: Luque A, Andreev V (eds) Concentrator photovoltaics, Chap 3. Springer, Berlin, pp 51–66
42. Bitnar B (2003) Silicon, germanium and silicon/germanium photocells for thermophotovoltaics applications. Semicond Sci Technol 18:221–227
43. Bett AW, Dimroth F, Stollwerck G, Sulima OV (1999) III-V compounds for solar cell applications. Appl Phys A A69(2):119–129
44. Bett AW, Sulima OV (2003) GaSb photovoltaic cells for applications in TPV generators. Semicond Sci Technol 18:184–190
45. Nagashima T, Corregidor V (2007) An overview of the contributions under cell technologies topic. Proceeding of the 7th world conference on the thermophotovoltaic

generation of electricity, Madrid, 25–27 Sept 2006. American Institute of Physics, pp 127–128
46. Dutta PS, Ostrogorsky AG, Gutmann RJ (1997) Bulk growth of GaSb and $Ga_{1-x}In_xSb$. Proceeding of the 3rd NREL Conference on thermophotovoltaic generation of electricity, Denver, Colorado, 18–21 May 1997. American Institute of Physics, pp 157–166
47. Dutta PS, Ostrogorsky AG, Gutmann RJ (1999) Bulk crystal growth of antimonide based III-V compounds for TPV applications. Proceeding of the 4th NREL Conference on thermophotovoltaic generation of electricity, Denver, Colorado, 11–14 Oct 1998. American Institute of Physics, pp 227–236
48. Vincent J, Díaz-Guerra C, Piqueras J, Diéguez E (2007) Technical developments and principal results of vertical feeding method for GaSb and GaInSb alloys. Proceeding of the 7th world Conference on thermophotovoltaic generation of electricity, Madrid, 25–27 Sept 2006. American Institute of Physics, pp 89–98
49. Alferov Zh I, Andreev VM, Rumyantsev VD (2007) III-V heterostructures in photovoltaics. In: Andreev V, Luque A (eds) Concentrator Photovoltaics, Chap 2. Springer, Berlin, pp 25–50
50. Sulima OV, Bett AW, Mauk MG, Mueller RL, Dutta PS, Ber BY (2003) GaSb-, InGaAsSb-, InGaSb- and InAsSbP TPV cells with Zn-diffused emitters. Proceeding of the 5th Conference on thermophotovoltaic generation of electricity, Rome, Italy, 16–19 Sept 2002. American Institute of Physics. pp 434–441
51. Wang CA (2004) Antimony-based III–V thermophotovoltaic materials and devices. Proceeding of the 6th international conference on thermophotovoltaic generation of electricity, Freiburg, Germany, 14–16 June 2004. American Institute of Physics, pp 255–266
52. Gevorkyan VA, Aroutiounian VM, Gambaryan KM, Arakelyan AH, Andreev IA, Golubev LV, Yakovlev YP, Wanlass MW (2007) The growth of low band-gap InAsSbP based diode heterostructures for thermo-photovoltaic application. Proceeding of the 7th world conference on thermophotovoltaic generation of electricity, Madrid, 25–27 Sept 2006. American Institute of Physics, pp 165–173
53. Horne WE, Morgan MD, Sundaram VS (1996) IR filters for TPV converter modules. Proceeding of the 2nd NREL conference on thermophotovoltaic generation of electricity, Colorado Springs, 16–20 July 1995. American Institute of Physics, pp 35–51
54. Fourspring PM, DePoy DM, Beausang JF, Gratrix EJ, Kristensen RT, Rahmlow TD, Talamo PJ, Lazo-Wasem JE, Wernsman B (2004) Thermophotovoltaic spectral control. Proceeding of the 6th international conference on thermophotovoltaic generation of electricity, Freiburg, Germany, 14–16 June 2004. American Institute of Physics, pp 171–179
55. Murthy SD, Langlois E, Bath I, Gutmann R, Brown E, Dzeindziel R, Freeman M, Choudhury N (1996) Characteristics of indium oxide plasma filters deposited by atmospheric pressure CVD. Proceeding of the 2nd NREL Conference on thermophotovoltaic generation of electricity, Colorado Springs, 16–20 July 1995. American Institute of Physics, pp 290–311
56. Fraas LM, Avery JE, Huang HX, Martinelli RU (2003) Thermophotovoltaic system configurations and spectral control. Semicond Sci Technol 18:165–173
57. Abbott P, Bett AW (2004) Cell-mounted spectral filters for thermophotovoltaic applications. Proceeding of the 6th International Conference on thermophotovoltaic generation of electricity, Freiburg, Germany, 14–16 June 2004. American Institute of Physics, pp 244–251
58. Green MA (2007) Single junction cells. In: Green MA (ed) Third generation photovoltaics advanced solar energy conversion, Chap 4. Springer, Berlin, pp 35–58
59. Wang CA, Murphy PG, OBrien PW, Shiau DA, Anderson AC, Liau ZL, Depoy DM, Nichols G (2003) Wafer-bonded internal back-surface reflectors for enhanced TPV performance. Proceeding of the 5th Conference on thermophotovoltaic generation of electricity, Rome, Italy, 16–19 Sept 2002. American Institute of Physics, pp 473–481

60. Volz W (2001) Entwicklung und Aufbau eines thermophotovoltaischen Energiewandlers (in German). Doctoral thesis, Universität Gesamthochschule Kassel, Institut für Solare Energieversorgungstechnik (ISET)
61. Charache GW, DePoy DM Baldasaro PF, Campbell BC (1996) Thermophotovoltaic device utilizing a back surface reflector for spectral control. Proceeding of the 2nd NREL Conference on Thermophotovoltaic Generation of Electricity, Colorado Springs, 16–20. July 1995. American Institute of Physics, pp 339–350
62. Fernández J, Dimroth F, Oliva E, Hermle M, Bett AW (2007) Back-surface Optimization of Germanium TPV Cells. Proceeding of the 7th world conference on thermophotovoltaic generation of electricity, Madrid, 25–27 Sept 2006. American Institute of Physics, pp 190–197
63. Nelson RE (2003) A brief history of thermophotovoltaic development. Semicond Sci Technol 18:141–143
64. Nelson RE (1995) Thermophotovoltaic emitter development. Proceeding of the 1st NREL Conference on Thermophotovoltaic Generation of Electricity, Copper Mountain, Colorado, 24–28 July 1994. American Institute of Physics, pp 80–96
65. Palfinger G, Bitnar B, Durisch W, Mayor J-C, Grützmacher D, Gobrecht J (2003) Cost estimate of electricity produced by TPV. Semicond Sci Technol 18:254–261
66. Qiu K, Hayden A (2004) A novel integrated TPV power generation system based on a cascaded radiant burner. Proceeding of the 6th International Conference on thermophotovoltaic generation of electricity, Freiburg, Germany, 14–16 June 2004. American Institute of Physics, pp 105–113
67. Swanson RM (2000) The promise of concentrators. Prog Photovoltaics Res Appl 8:93–111
68. Hampe C, Metz A, Hezel R (2002) Innovative Silicon-concentrator solar cell for thermophotovoltaic application. Proceeding of the 17th european photovoltaic solar energy conference, Munich, 22–26 Oct 2001. WIP, pp 18–22
69. Posthuma N, Heide J, Flamand G, Poortmans J (2004) Development of low cost germanium photovoltaic cells for application in TPV using spin-on diffusants. Proceeding of the 6th International Conference on Thermophotovoltaic Generation of Electricity, Freiburg, Germany 14–16 June 2004. American Institute of Physics, pp 337–344
70. Nagashima T, Okumura K, Yamaguchi M (2007) A germanium back contact type thermophotovoltaic cell. Proceeding of the 7th World Conference on the thermophotovoltaic generation of electricity, Madrid, 25–27 Sept 2006. American Institute of Physics, pp 174–181
71. Palfinger G (2006) Low dimensional Si/SiGe structures deposited by UHV-CVD for thermophotovoltaics. Doctoral thesis, Paul Scherrer Institut
72. Yamaguchi M (2001) Super-high efficiency III-V tandem and multijunction cells. In: Archer MD, Hill R (eds) Clean electricity from photovoltaics, Chap 8, vol 1. Series on photoconversion of solar energy, Imperial College Press, London
73. Fraas L, Ballantyne R, She-Hui Shi-Zhong Ye, Gregory S, Keyes J, Avery J Lamson D, Daniels B (1999) Commercial GaSb cell and circuit development for the Midnight Sun(R) TPV stove. Proceeding of the 4th NREL Conference on thermophotovoltaic generation of electricity, Denver, Colorado, 11–14 Oct 1998. American Institute of Physics, pp 480–487
74. Corregidor V, Vincent J, Algora C, Diéguez E (2007) Thermophotovoltaic converters based on poly-crystalline GaSb. Proceeding of the 7th world conference on thermophotovoltaic generation of electricity, Madrid, 25–27 Sept 2006. American Institute of Physics, pp 157–164
75. Contreras M, Wiesner H, Webb J (1997) Thin-film polycrystalline $Ga_{1-x}In_xSb$ materials. Proceeding of the 3rd NREL conference on thermophotovoltaic generation of electricity, Denver, Colorado, 18–21 May 1997. American Institute of Physics, pp 403–410
76. Dhere NG (1997) Appropriate materials and preparation techniques for polycrystalline-thin-film thermophotovoltaic cells. Proceeding of the 3rd NREL conference on thermophotovoltaic generation of electricity, Denver, Colorado, 18–21 May 1997. American Institute of Physics, pp 423–442

77. Wanlass MW, Schwartz RJ (1995) Introduction to workshop spectral control and converters. Proceeding of the 1st NREL conference on thermophotovoltaic generation of electricity, Copper Mountain, Colorado, 24–28 July 1994. American Institute of Physics, pp 6–12
78. Zheng L, Sweileh GM, Haywood SK, Scott CG, Lakrimi M, Mason NJ, Walker PJ (1999) p-GaSb/n-GaAs heterojunctions for thermophotovoltaic cells grown by MOVPE. Proceeding of the 4th NREL conference on thermophotovoltaic generation of electricity, Denver, Colorado, 11–14 Oct 1998. American Institute of Physics, pp 525–534
79. Wanlass MW, Ahrenkiel SP, Ahrenkiel RK, Carapella JJ, Wehrer RJ, Wernsman B (2004) Recent advances in low-bandgap, InP-Based GaInAs/InAsP materials and devices for thermophotovoltaic (TPV) energy conversion. Proceeding of the 6th international conference on thermophotovoltaic generation of electricity, Freiburg, Germany, 14–16 June 2004. American Institute of Physics, pp 427–435
80. Fraas LM, Avery JE, Sundaram VS, Dinh VT, Davenport TM, Yerkes JW, Gee JM, Emery KA (1990) Over 35% efficient GaAs/GaSb stacked concentrator cell assemblies for terrestrial applications. Proceeding of the 21st IEEE photovoltaic specialists conference, pp 190–195
81. Khvostikov VP, Gazaryan PY, Khvostikova OA, Potapovich NS, Sorokina, SV, Malevskaya AV, Shvarts MZ, Shmidt NM, Andreev VM (2007) GaSb Applications for solar thermophotovoltaic conversion. Proceeding of the 7th world conference on thermophotovoltaic generation of electricity, Madrid, 25–27 Sept 2006. American Institute of Physics, pp 139–148
82. Wojtczuk S, Gagnon E, Geoffroy L, Parodos T (1995) $In_xGa_{1-x}As$ thermophotovoltaic cell performance vs. bandgap. Proceeding of the 1st NREL conference on thermophotovoltaic generation of electricity, Copper Mountain, Colorado, 24–28 July 1994. American Institute of Physics, pp 177–187
83. Karlina LB, Blagnov PA, Kulagina MM, Vlasov AS, Vargas-Aburto C, Uribe RM (2003) Zinc (P) diffusion in $In_{0.53}Ga_{0.47}As$ and GaSb for TPV devices. Proceeding of the 5th conference on thermophotovoltaic generation of electricity, Rome, Italy, 16–19 Sept 2002. American Institute of Physics, pp 373–382
84. Ginige R, Kelleher C, Corbett B, Hilgarth J, Clarke G (2002) The design, fabrication and evaluation of InGaAs/InP TPV cells for commercial applications. Proceeding of the 5th conference on thermophotovoltaic generation of electricity, Rome, Italy, 16–19 Sept 2002. American Institute of Physics, pp 354–362
85. Murray SL, Newman FD, Murray CS, Wilt DM, Wanlass MW, Ahrenkiel P, Messham R, Siergiej RR (2003) MOCVD growth of lattice-matched and mismatched InGaAs materials for thermophotovoltaic energy conversion. Semicond Sci Technol 18:202–208
86. Wanlass MW, Carapella JJ, Duda A, Emery K, Gedvilas L, Moriarty T, Ward S, Webb J, Wu X, Murray CS (1999) High-Performance, 0.6 eV, GaInAs InAsP thermophotovoltaic converters and monolithically interconnected modules. Proceeding of the 4th NREL conference on thermophotovoltaic generation of electricity, Denver, Colorado, 11–14 Oct 1998. American Institute of Physics, pp 132–141
87. Wilt D, Wehrer R, Palmisiano M, Wanlass M, Murray C (2003) Monolithic interconnected modules (MIMs) for thermophotovoltaic energy conversion. Semicond Sci Technol 18:209–215
88. Siergiej RR, Wernsman B, Derry SA, Wehrer RJ, Link SD, Palmisiano, MN, Messham RL, Murray S, Murray CS, Newman F, Hills J, Taylor D (2003) 20% efficient InGaAs/InPAs thermophotovoltaic cells. Proceeding of the 5th conference on thermophotovoltaic generation of electricity, Rome, Italy, 16–19 Sept 2002. American Institute of Physics, pp 414–423
89. Ringel SA, Sacks RN, Qin L, Clevenger MB, Murray CS (1999) Growth and properties of InGaAs/FeAl/InAlAs/InP heterostructures for buried reflector/interconnect applications in InGaAs thermophotovoltaic devices. Proceeding of the 4th NREL conference on thermophotovoltaic generation of electricity, Denver, Colorado, 11–14 Oct 1998. American Institute of Physics, pp 142–151

90. Ward JS, Duda A, Wanlass MW, Carapella JJ, Wu X, Matson RJ, Coutts TJ, Moriarty T, Murray CS, Riley DR (1997) Novel design for monolithic interconnected modules (MIMS) for thermophotovoltaic power conversion. Proceeding of the 3rd NREL conference on thermophotovoltaic generation of electricity, Denver, Colorado, 18–21 May 1997. American Institute of Physics, pp 227–236
91. Wernsman B, Siergiej RR, Link SD, Mahorter RG, Palmisiano MN, Wehrer RJ, Schultz RW, Schmuck GP, Messham RL, Murray S, Murray CS, Newman F, Taylor D, DePoy DM, Rahmlow T (2004) Greater than 20% radiant heat conversion efficiency of a thermophotovoltaic radiator/module system using reflective spectral control. Trans Electron Devices 51(3):512–515
92. Shellenbarger Z, Taylor G, Martinelli R, Carpinelli J (2004) High performance InGaAsSb thermophotovoltaic cells via multi-wafer OMVPE growth. Proceeding of the 6th International Conference on thermophotovoltaic generation of electricity, Freiburg, Germany, 14–16 June 2004. American Institute of Physics, pp 314–323
93. Mauk MG, Shellenbarger ZA, Gottfried MI, Cox JA, Feyock BW, McNeely JB, DiNetta LC, Mueller RL (1997) New concepts for III-V antimonide thermophotovoltaics. Proceeding of the 3rd NREL conference on thermophotovoltaic generation of electricity, Denver, Colorado, 18–21 May 1997. American Institute of Physics, pp 129–137
94. Imenes AG, Mills DR (2004) Spectral beam splitting technology for increased conversion efficiency in solar concentrating systems: a review. Sol Energy Mater Sol Cells 84:19–69
95. Green MA (2003) Tandem cells. In: Green MA (ed) "Third generation photovoltaics - advanced solar energy conversion", Chap 5. Springer, Berlin, pp 59–67
96. Jackson ED (1955) Areas for Improvement of the semiconductor solar energy converter. Trans Conf Use Sol Energy 5:122–126, 31 Oct – 1 Nov
97. Algora C (2007) Very-High-concentration challenges of III-V multijunction solar cells. In: Luque A, Andreev V (eds) Concentrator Photovoltaics, Chap 5. Springer, Berlin, pp 89–111
98. Rumyantsev VD, Khvostikov VP, Sorokina O, Vasilev AI, Andreev VM (1999) Portable TPV generator based on metallic emitter and 1.5-amp GaSb cells. Proceeding of the 4th NREL Conference on Thermophotovoltaic Generation of Electricity, Denver, Colorado, 11–14 Oct 1998. American Institute of Physics, pp 384–393
99. Wilt DM, Wehrer RJ, Maurer WF, Jenkins PP, Wernsman B, Schultz RW (2004) Buffer layer effects on tandem InGaAs TPV devices. Proceeding of the 6th international conference on thermophotovoltaic generation of electricity, Freiburg, Germany 14–16 June 2004. American Institute of Physics, pp 453–461
100. Siergiej RR, Sinharoy S, Valko T, Wehrer RJ, Wernsman B, Link SD, Schultz RW, Messham RL (2004) InGaAsP/InGaAs tandem TPV device. Proceeding of the 6th international conference on thermophotovoltaic generation of electricity, Freiburg, Germany, 14–16 June 2004. American Institute of Physics, pp 480–488
101. Rohr C (2000) InGaAsP quantum well cells for thermophotovoltaic applications. Imperial college of science technology and medicine, London
102. Connolly JP, Rohr C (2003) Quantum well cells for thermophotovoltaics. Semicond Sci Technol 18:216–220
103. Hardingham CM (2001) Chapter 13: cells and systems for space applications. In: Archer MD, Hill R (eds) Clean electricity from photovoltaics, vol 1. Series on photoconversion of solar energyImperial College Press, London
104. Sater BL (1995) Vertical multi-junction cells for thermophotovoltaic conversion. Proceeding of the 1st NREL Conference on Thermophotovoltaic Generation of Electricity, Copper Mountain, Colorado, 24–28 July 1994. American Institute of Physics, pp 165–176
105. Luque A (2007) Solar Thermophotovoltaics: Combining solar thermal and photovoltaics. Proceeding of the 7th world conference on thermophotovoltaic generation of electricity, Madrid, 25–27 Sept 2006. American Institute of Physics, pp 3–16
106. Chen Z, Brandhorst HW (1999) Effect of elevated temperatures on the performance of an InP cell illuminated by a selective emitter. Proceeding of the 4th NREL conference on

thermophotovoltaic generation of electricity, Denver, Colorado, 11–14 Oct 1998. American Institute of Physics, pp 438–445
107. Chen Z, Brandhorst HW, Wells BK (2001) InAsP cells for solar thermophotovoltaic applications. IEEE Aerosp Electron Syst Mag 16(4):39–43
108. Brandhorst HW, Chen Z (2000) Thermophotovoltaic conversion using selective infrared line emitters and large band gap photovoltaic devices, Auburn University, US Patent 6072116
109. Luque A, Martí A, Cuadra L, Algora C, Wahnon P, Sala G, Benítez P, Bett AW, Gombert A, Andreev VM, Jassaud C, Van Roosmalen JAM, Alonso J, Räuber A, Strobel G, Stolz W, Bitnar B, Stanley C, Conesa JC, Van Sark W, Barnham K, Danz R, Meyer T, Luque-Heredia I, Kenny R, Christofides C (2004) FULLSPECTRUM: a new PV wave making more efficient use of the solar spectrum. Proceeding of the 19th European photovoltaic solar energy conference and exhibition, Paris, 7–11 June
110. Corkish R, Green MA, Puzzer T (2002) Solar energy collection by antennas. Sol Energy 73:395–401
111. Goswami DY, Vijayaraghavan S, Lu S, Tamm G (2004) New and emerging developments in solar energy. Sol Energy 76:33–43
112. Sala G, Antón I, Domínguez C (2007) Qualification testing of TPV systems and components: first steps. Proceeding of the 7th world conference on thermophotovoltaic generation of electricity, Madrid, 25–27 Sept 2006. American Institute of Physics, pp 251–261
113. Gethers CK, Ballinger CT, DePoy DM (1999) Lessons learned on closed cavity TPV system efficiency measurement. Proceeding of the 4th NREL conference on thermophotovoltaic generation of electricity, Denver, Colorado, 11–14 Oct 1998. American Institute of Physics, pp 335–348
114. Gethers CK, Ballinger CT, Postlethwait MA, DePoy DM, Baldasaro PF (1997) TPV efficiency predictions and measurements for a closed cavity geometry. Proceeding of the 3rd NREL conference on thermophotovoltaic generation of electricity, Denver, Colorado, 18–21 May 1997. American Institute of Physics, pp 471–486
115. Emery K (2003) Characterizing thermophotovoltaic cells. Semicond Sci Technol 18:228–231
116. Emery K, Basore P (1995) Workshop: device and system characterization: consensus recommendations. Proceeding of the 1st. NREL conference on thermophotovoltaic generation of electricity, Copper Mountain, Colorado, 24–28 July 1994. American Institute of Physics, pp 23–26
117. Burger DR, Mueller RL (1995) Characterization of thermophotovoltaic cells. Proceeding of the 1st NREL conference on thermophotovoltaic generation of electricity, Copper Mountain, Colorado, 24–28 July 1994. American Institute of Physics, pp 457–72
118. Zierak M, Borrego J, Bhat I, Ehsani H, Marcy D, Gutmann R, Parrington J, Charache G, Nichols G (1995) Characterization of InGaAs TPV cells. Proceeding of the 1st NREL conference on thermophotovoltaic generation of electricity, Copper Mountain, Colorado, 24–28 July 1994. American Institute of Physics, pp 473–483
119. Mattarolo G (2007) Development and modelling of a thermophotovoltaic system, Doctoral thesis, University of Kassel
120. Durisch W, Grob B, Mayor JC, Panitz JC, Rosselet A (1999) Interfacing a small thermophotovoltaic generator to the grid, 4 NREL conference on thermophotovoltaic generation of electricity, Denver, Colorado, 11–14 Oct 1998. American Institute of Physics, pp 403–414
121. Doyle EF, Becker FE, Shukla KC, Fraas LM (1999) Design of a thermophotovoltaic battery substitute. Proceeding of the 4th NREL Conference on thermophotovoltaic generation of electricity, Denver, Colorado, 11–14 Oct 1998. American Institute of Physics, pp 351–361
122. Carlson RS, Fraas LM (2007) Adapting TPV for use in a standard home heating furnace. Proceeding of the 7th world conference on thermophotovoltaic generation of electricity, Madrid, 25–27 Sept 2006. American Institute of Physics, pp 273–279

Part II
Systems

Chapter 5
Heat Transfer Theory and System Modelling

5.1 Introduction

This chapter discusses some heat transfer theory relevant to TPV conversion. Heat transfer is concerned with the transport of thermal energy due to a temperature difference. Time independent (steady-state) and time dependent (transient) heat transfer can be distinguished. For TPV conversion in most cases steady-state heat transfer is considered. Start-up and shutdown processes of the TPV system may require addressing transient heat transfer and thermo-mechanical stress phenomena. Three modes of heat transfer are distinguished namely, conduction, convection and radiation. Figure 5.1 shows schematically the arrangement of a typical TPV system. The system consists of a heat source thermally connected to the radiator and a heat sink thermally connected to the PV cells, as well as thermal insulation with a reflective surface. Radiator, reflective thermal insulation and PV cell form the TPV cavity. Such arrangement can be used to explain the relevant heat transfer modes for TPV conversion.

Both conductive and convective heat transfer from the radiator to the PV cell are generally parasitic and need to be minimised. Losses from the heat source via the thermal insulation to the surroundings need to be also minimised. Ideally, only low-grade heat from the heat sink would be transferred to the surroundings. In contrast to this, radiative heat transfer in the cavity is desired. Radiation needs to be optimised in terms of the spectral, angular and spatial radiation distribution in order to achieve high electrical power densities and efficiencies. It needs to be stressed that the angular and spatial radiation distribution differs fundamentally from the collimated solar radiation. An ideal blackbody radiator emits angularly with a lambertian distribution radiation and usually reflective thermal insulation areas are required to reduce cavity losses. It is generally desirable to minimise the reflective insulation area to avoid losses via this path (see view factors in Sect. 5.4.4).

In terms of the radiative heat transfer spectrum, the figures of merit can be defined as high in-band radiation and low out-of-band radiative heat transfer.

T. Bauer, *Thermophotovoltaics*, Green Energy and Technology,
DOI: 10.1007/978-3-642-19965-3_5,

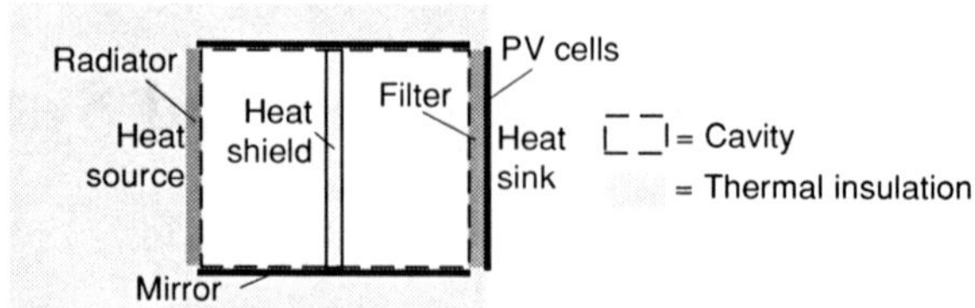

Fig. 5.1 Schematic of a typical TPV system with cavity

In other words, radiative heat transfer from the emitter to cell for photons with energies $h\nu$ larger than the cell bandgap $h\nu_g$ should be maximised (high in-band radiative heat transfer) and heat transfer of photons with $h\nu < h\nu_g$ should be minimised (low out-of-band radiative heat transfer).

Exact modelling of heat transfer in TPV systems is generally complex. For example, radiative surface properties of cavity components depend on the temperature, angle and spectrum and these components can interact by the three heat transfer modes (conduction, convection and radiation). Hence, for heat transfer modelling, simplifications need to be made. For example, modelling can be limited to the dominating heat transfer mode, surfaces can be approximated as either diffuse or specularly reflective in terms of their angular dependence and spectral dependences can be simplified by assuming constant values within spectral bands.

Sections 5.2–5.4 discuss the theory and relevance to TPV conversion of the three heat transfer modes with the emphasis on radiation. Section 5.5 focus on the theory of coupled radiative and conductive heat transfer in semitransparent media (e.g., glass).

5.2 Conduction

In the process of heat conduction, energy is transferred between neighbouring molecules, where the mean position of the molecule is fixed. For steady-state conditions and a known thermal resistance R_{th} (unit K/W), the heat transfer rate P (unit W) from an isothermal surface at T_h to another isothermal surface with temperature T_c can be calculated by Eq. 5.1. The thermal resistance can be calculated by the shape factor S (unit meter) assuming a homogeneous isotropic material between the two isothermal surfaces with a thermal conductivity k that is independent of the temperature. For a large number of geometries the shape factor S can be calculated. Equation 5.2 defines S for two isothermal rectangular parallel plates with equal size L_1 times L_2 and with the plate distance L assuming that there is no heat transfer on the remaining open surfaces (ideal insulation, or adiabatic boundary condition). Equation 5.3 defines S for two concentric tubes with an equal length of L and adiabatic boundary conditions, where the inner tube has the outer radius r_{small} and the outer tube the inner radius r_{large}. The geometries defined in Eqs. 5.2 and 5.3 are illustrative examples. These equations are often useful to estimate approximately the conduction heat transfer in a thermal system. In the literature a lot more shape factors can be found [1, 2].

Table 5.1 Overview of areas with conduction heat transfer in a TPV system

Undesirable conduction heat transfer	Utilisation of (high) conduction heat transfer
Heat transfer via gas conduction from the radiator to the PV cell	High heat transfer in the radiator to achieve a uniform radiator surface temperature (leads to a uniform radiation)
Heat transfer via the (reflective) thermal insulation from the radiator to the PV cell	Low thermal resistance for the PV cell-to-sink interconnection (Sect. 4.7.4) [3]
Heat losses from the system (e.g., radiator, heat shield) via the thermal insulation to the surrounding (see Sect. 6.4)	Combustion system: high conductive recuperator for fuel/air preheating from flue gas

$$P = \frac{T_h - T_c}{R_{th}} = kS(T_h - T_c) \quad (5.1)$$

$$S_{Plate} = \frac{L_1 L_2}{L} \quad (5.2)$$

$$S_{conc.tube} = \frac{2\pi}{\ln\left(\frac{r_{large}}{r_{small}}\right)} L \quad (5.3)$$

Conduction heat transfer can occur in different areas in a TPV system. In some cases it is desirable and in other cases conduction heat transfer needs to be minimised (Table 5.1).

5.3 Convection

In the process of heat convection, macroscopic movement of matter transfers energy. It is the transport of energy to or from a surface by the combined action of conduction heat transfer and fluid motion. Convection heat transfer at solid-fluid surfaces is most commonly of interest, but fluid-fluid interfaces are also encountered. It can be distinguished between *forced* convection and *free* convection. Different densities of matter at different temperatures cause free (or natural) convection. Forced convective heat transfer involves fluids that are moved by some mechanical means [1]. The fluids can have either a single aggregate state (single-phase fluid) or show a phase change (two-phase fluid) between liquid and gas phase.

Convective heat transfer processes are usually characterised in terms of a convective heat transfer coefficient h with the unit W/(m^2 K), where H is the heat transfer rate (unit W/m^2) and T_w is the wall temperature and T_F the fluid temperature (Eq. 5.4). Table 5.2 summarises convection heat transfer processes that can occur in a TPV system.

$$H = h(T_w - T_F) \quad (5.4)$$

Table 5.2 Overview of areas with convection heat transfer in a TPV system

Undesirable convection heat transfer	Utilisation of (high) convection heat transfer
Parasitic free convection of the gas inside the TPV cavity (e.g., inert gas)	PV cell cooling (Sect. 4.7.4):
Convective boundary conditions of the entire insulated system to the surrounding	Forced convection, single-phase fluid
	Forced convection, two-phase fluid
	Free convection (thermo-syphon with water)
	Combustion system:
	Recuperator fuel/air preheating from flue gas
	Heat transfer from combustion zone to radiator

5.4 Radiation

The importance of radiative heat transfer modelling for TPV conversion has been highlighted by White and Hottel [4]. Subsequent TPV heat transfer modelling was summarised by Coutts [5] and Aschaber et al. [6–8]. In the following the relevant radiation theory is discussed. A fundamental treatment of radiative heat transfer can be found in the literature [9–12].

All materials continuously emit or absorb electromagnetic waves, or photons, by lowering or raising their molecular energy levels. Heat transfer by radiation does not require a medium unlike heat transfer by conduction and convection [9, 10]. Hence, unlike conduction and convection, radiation heat transfer also occurs in a vacuum system.

The quantum theory describes radiation as single photon with the energy $h\nu$, where h is Planck's constant and v is the frequency of the photon. The theory of electromagnetic waves states that radiation has the velocity of the speed of light c_0 in vacuum. In a medium with refractive index n the speed is reduced to the speed c Eq. 5.5. On the other hand, the frequency, or energy, of the photon remains unchanged. Equation 5.6 describe the relation between wavelength λ and the frequency ν [2]. In other literature, sometimes the wave number, defined as the reciprocal of the wavelength in vacuum, is also used [10].

$$n = \frac{c_0}{c} \tag{5.5}$$

$$\lambda \cdot \nu = c \tag{5.6}$$

5.4.1 Absorption of Radiation

When radiation travelling through a medium strikes the surface of a second medium, radiation may be reflected (either partly or totally), and any non-reflected part will penetrate into the second medium [10]. While passing into the second medium, the *Lambert-Beer law* describes the radiation reduction assuming 1D radiative heat transfer, a non-scattering medium and a constant absorption coefficient $\alpha(\lambda)$ along a path length S, where $I_{0,\lambda}$ is the initial radiation intensity and $I_{L,\lambda}$ the radiation intensity after a path length S (Eq. 5.7).

$$I_{L,\lambda} = I_{0,\lambda} \cdot e^{-\alpha(\lambda) \cdot S} \tag{5.7}$$

As discussed in Sect. 2.4.3, three cases for the optical thickness $\alpha(\lambda) \cdot S$ can be distinguished [9]:

- Opaque or optically thick: $\alpha(\lambda) \cdot S \gg 1$.
- Transparent or optically thin: $\alpha(\lambda) \cdot S \ll 1$.
- Semitransparent: for all other cases of $\alpha(\lambda) \cdot S$.

Metals tend to require a small path length S to become opaque, although very thin metal layers behave as transparent for some wavelengths. For TPV, gold layers with a thickness from 15 to 25 nm have been used within dielectric filters [13]. Non-metals generally require a much larger thickness before they become opaque [10]. For an opaque medium, although radiation penetrates into the medium, interaction with radiation can be treated as a surface phenomenon. For semitransparent media, there are radiation interactions with the surface and within the medium. Examples in a TPV system are filament ceramic radiators, glass shields and PV cells.

5.4.2 Emission of Radiation

The upper limit of radiation emitted at every wavelength λ (or frequency v) and in every direction into a medium with refractive index n is given by a blackbody as described by Planck's function. The function depends strongly on the radiator temperature T_s. Equation 5.8 shows Planck's function depending on either the wavelength λ (top) or the frequency v (bottom), where the refractive index n is assumed wavelength and temperature independent [9–11].

$$\begin{aligned} i_{b\lambda(\lambda,T_s)} &= \frac{2hc_0^2}{n^2\lambda^5} \cdot \frac{1}{e^{\frac{hc_0}{kT_s n\lambda}} - 1} \\ i_{bv(v,T_s)} &= \frac{2n^2hv^3}{c_0^2} \cdot \frac{1}{e^{\frac{hv}{kT_s}} - 1} \end{aligned} \tag{5.8}$$

Alternatively, Planck's function can be written in a normalised form depending on a newly defined variable x (or $\tilde{X}$) (Eq. 5.9). It can be seen that a change in radiator temperature T_s does not change the shape of i_b, but scales the normalised function equally for all values of x (or $\tilde{X}$). In other words, the *shape* of the function depends solely on the newly defined variable x (or $\tilde{X}$) instead of the two variables wave length λ (or frequency v) and radiator temperature T_s.

$$
\begin{aligned}
i_{b\lambda(\tilde{x},T_s)} &= \frac{2n^3k^5T_s^5}{h^4c_0^3}\cdot\frac{\tilde{x}^5}{e^{\tilde{x}}-1} \quad \text{with} \quad \tilde{x}=\frac{hc_0}{kT_sn\lambda} \\
i_{bv(x,T_s)} &= \frac{2n^2h}{c_0^2}\left(\frac{kT_s}{h}\right)^3\frac{x^3}{e^x-1} \quad \text{with} \quad x=\frac{hv}{kT_s}
\end{aligned} \tag{5.9}
$$

From Eq. 5.9 Wien's displacement law can be derived. The maximum of the Eq. 5.9 depends on the single variable x (or $\tilde{X}$) and is given by Eq. 5.10.

$$
\begin{aligned}
\frac{\mathrm{d}}{\mathrm{d}x}\left(\frac{\tilde{x}^5}{e^{\tilde{x}}-1}\right) &= 0 \rightarrow \tilde{x}_{\max} = 4.9651 \\
\frac{\mathrm{d}}{\mathrm{d}x}\left(\frac{x^3}{e^x-1}\right) &= 0 \rightarrow x_{\max} = 2.8214
\end{aligned} \tag{5.10}
$$

The wave length (or frequency) with the maximum of the function $i_{b\lambda(\lambda,Ts)}$ and $i_{bv(v,Ts)}$ can be obtained, if the value x_{max} of Eq. (5.10) and the definition of x (or$\tilde{X}$) is used Eq. 5.11 [10].

$$
\begin{aligned}
\lambda_{\max} &= \frac{hc_0}{4.9651\cdot kT_sn} = \frac{b}{n\cdot T_s} \quad \text{with} \quad \mathrm{b} = 2898\ \mu\mathrm{m\ K} \\
v_{\max} &= \frac{2.8214kT_s}{h}
\end{aligned} \tag{5.11}
$$

Figure 5.2 illustrates the wavelength shift of the blackbody maximum with its temperature described by Wien's displacement law. It can be seen that higher radiator temperatures result in a shift of radiation towards shorter wavelengths (or higher frequencies). For TPV conversion, the implication of this law is that systems with a higher radiator temperature can utilise PV cell with a higher energy bandgap. For example a silicon PV cell with a bandgap energy of $hv_g = 1.12$ eV (responds up to 1.11 μm) could be matched to the maximum of the blackbody at 2,610 K. It is interesting to note that exactly one-fourth of the total radiation lies in the wavelength range below the peak wavelength regardless of the radiator temperature [10]. For the silicon cell example this means that even though a very high radiator temperature of 2,610 K was selected, there are still three-quarter of the total radiation in the long wave length range that cannot be converted! This shows the vital importance to avoid, or spectrally control, long wavelength radiative heat transfer from the radiator to the PV cell. Figure 5.2 shows additionally the solar spectrum with a significantly higher portion of radiation in the visible spectral range (0.38–0.76 μm). The maximum wavelength in Fig. 5.2 is 4 μm in order to show the solar spectrum in detail. However, it should be considered that significant radiative heat transfer occurs up to 20 μm. The wavelength range of interest was already discussed (see Table 3.1).

Integration of $i_{b\lambda\ (\lambda,Ts)}$ (Eq. 5.8) over all wavelengths results in the *total blackbody intensity* per solid angle (Eq. 5.12) [9, 10]. Equation 5.13 defines the Stefan-Boltzmann constant. The total emitted radiation depends strongly on the temperature and increases with the fourth power. As a result of this law, a higher

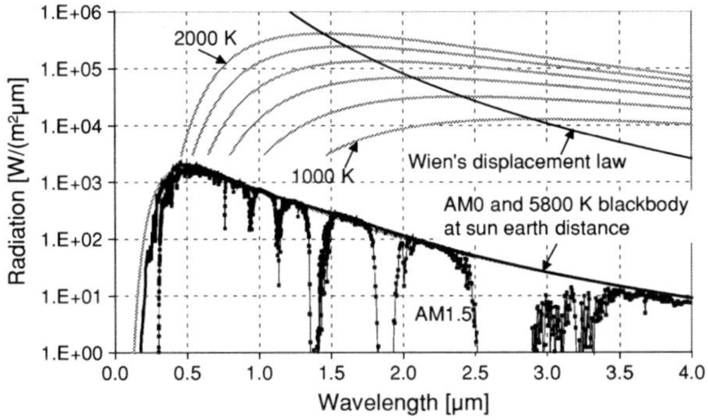

Fig. 5.2 Comparison of solar spectra at sun earth distance and blackbody spectra in semi-logarithmic scale. The blackbody radiation increases from 1,000 to 2,000 K in 200 K steps (small values overlapping with the solar spectra are not shown). The blackbody maximum given by Wien's displacement law is also shown (*black*). The AM1.5 solar spectrum (*black*) shows strong absorption bands, whereas the AM0 spectrum (*black*) closely matches a 5,800 K blackbody at sun earth distance (*grey*)

radiator temperature results in higher radiative heat transfer to the cell and this usually leads to higher electrical power densities expressed in the unit W/cm^2.

$$i_{\mathrm{b}(T_s)} = \int_0^\infty i_{\mathrm{b}\lambda(\lambda,T_s)} \mathrm{d}\lambda = \frac{n^2 \sigma T_s^4}{\pi} \tag{5.12}$$

$$\sigma = \frac{2\pi^5 k^4}{15 h^3 c_0^2} = 5.670 \cdot 10^{-8} \frac{W}{m^2 K^4} \tag{5.13}$$

5.4.3 Radiation Interaction at Surfaces

Blackbody surfaces are by definition diffuse, so that the directionally emitted radiation decreases with polar angle θ away from normal, which is known as the *Lambert's cosine-law* (Eq. 5.14) [9]. Assuming a TPV system consisting of two infinite plates, one a blackbody radiator and the other one a PV cell, it becomes clear that radiation is incident *at all angles at the PV cell*, whereas the cell surface can only absorb radiation under certain angles. This angular radiation distribution is fundamentally different from solar radiation. For direct solar radiation on the earth, the sun is approximately a point source and radiation rays are parallel with a defined incident angle, if diffuse solar radiation is not considered.

$$I_{\mathrm{b}\lambda(\lambda,T_s,\theta)} = i_{b\lambda(\lambda,T_s)} \cdot \cos(\theta) \tag{5.14}$$

Integration of Planck's function Eq. 5.8 over one hemisphere leads to *Planck's radiation law* (Eq. 5.15), which defines radiation emitted from a blackbody surface into a second medium with refractive index *n* over one hemisphere. For TPV conversion, most commonly this second medium is a gas and the refractive index is in good approximation unity.

$$I_{b\lambda(\lambda,T_s)} = n^2 \cdot \pi \cdot i_{b\lambda(\lambda,T_s)}$$

$$I_{bv(v,T_s)} = n^2 \cdot \pi \cdot i_{bv(v,T_s)} \tag{5.15}$$

Integration over all wave lengths λ of Eq. 5.15 results in the total hemispherical radiated power $I_{b(Ts)}$, which is known as the *Stefan-Boltzmann law* Eq. 5.16.

$$I_{b(T_s)} = \int_0^\infty I_{bv(v,T_s)} \quad dv = n^2 \sigma T_s^4$$

$$I_{b(T_s)} = \int_0^\infty I_{b\lambda(\lambda,T_s)} \quad d\lambda = n^2 \sigma T_s^4 \tag{5.16}$$

The *emissivity* is then defined as the ratio of real to blackbody radiation varying from 0 to 1, where the definition can be directional spectral (e.g., commonly for normal direction), directional total, hemispherical spectral or hemispherical total [9]. Additionally the surface emissivity depends on temperature, material composition (e.g., oxide films) and physical structure (e.g., rough, polished or engineered microstructures). For example selective tungsten radiators with surface structures in the same order of dimension as the radiation wavelength have been fabricated. Surfaces with spectrally constant emissivity are known as grey surfaces. In order to simplify the modelling, spectral bands can be defined with a constant emissivity, or grey characteristic within a spectral band. The surface can be not only engineered spectrally but also to be directionally selective [10]. This could be useful for directionally selective emitting TPV radiators, since PV cell surfaces become increasingly reflective for angles away from normal. So far, the work on directionally selective radiators is limited in the TPV community. Les et al. discussed this approach [14].

The *absorptivity* is defined as the ratio of absorbed to incident radiation, where the definition may be directional or hemispherical, as well as spectral or total [10]. Often Kirchhoff's law is valid, which states that the absorptivity is equal to the emissivity. The conditions for the correctness of this law are discussed elsewhere [9, 10].

The *reflectivity* of a surface depends not only on the direction of the incoming radiation, as for the absorptivity, but also on the direction of the reflected radiation. Two ideal cases of reflection can be defined. These are diffuse (equal reflection in all directions) and specular (mirror like). Specularly reflective surfaces in a TPV system can include mirrors, PV cells and glass surfaces. Real surface reflection can be often approximated by specular, diffuse or combined specular-diffuse reflection.

The *radiative heat transfer* between surfaces can be calculated using view factors. For two infinite parallel plates with blackbody surface characteristics (view factor 1, emissivity 1) the radiative heat flux H (W/m^2) from the hot to the cold plate is given by Eq. 5.17, assuming no other modes of heat transfer (conduction and convection) and a transparent medium in between the surfaces [10]. This equation can be regard as the simplest model to describe radiative heat transfer from the radiator to the PV cell as discussed in the next section.

$$H = n^2\sigma \cdot \left(T_{\rm h}^4 - T_{\rm c}^4\right) \tag{5.17}$$

5.4.4 Radiative Heat Transfer within TPV Cavities

Radiative heat transfer between two surfaces: Eq. 5.18 gives the net radiative heat transfer H (unit W/m^2) from a hot plate to a cold plate. This equation assumes two isothermal blackbody surfaces with the temperature $T_{\rm h}$ and $T_{\rm c}$ and no participating medium ($n = 1$) between the two surfaces [9–11]. The view factor $F_{\rm h\text{-}c}$ depends on the arrangement of the plates (distance, angle and size) and reaches unity for an infinite parallel plate arrangement. The view factor may be also known as configuration factor, radiation shape factor or angle factor. View factors for a variety of geometries can be found in the literature [9, 10].

$$H = \sigma \cdot \left(T_{\rm h}^4 - T_{\rm c}^4\right) \cdot F_{\rm h-c} \tag{5.18}$$

Equation 5.18 allows simplified radiative heat transfer calculations from the radiator with temperature $T_{\rm h}$ to the PV cell with the temperature $T_{\rm c}$. For the temperature of interest in TPV, the term $T_{\rm h}^4$ is usually much larger compared to $T_{\rm c}^4$, so that the latter term can be in a first approximation neglected. The equation can give an upper limit of the total radiative heat transfer rate with the unit W/cm^2. For example, $T_{\rm h} = 1{,}000$ K results in a total radiative flux of 5.6 W/cm^2 and $T_{\rm h} = 2{,}000$ K in 91 W/cm^2 assuming $T_{\rm c} = 300$ K.

For analytical TPV modelling, Eq. 5.18 has been the basis. Hottel computed the radiative heat transfer through glass shields for an infinite plate arrangement using an analytical model. He included spectral dependences of the radiator emissivity (ε) and the reflectivity of PV cell and glass shield. Glass shield absorption was assumed to be wavelength independent and participation of the glass was neglected. The reflectance for different angles was assumed constant. One conclusion made was that the shield reduces in-band transmission [4]. Burger [15] and Schroeder et al. [16] developed their own models based on the Hottel model.

Net radiation method for a TPV cavity: A TPV cavity is often composed of a radiator, PV cells and reflector surfaces as shown in Fig. 5.1 [6, 17]. In some cases the assumption can be made that the medium within the enclosure shows no interaction with radiation and is non-conductive and non-convective (e.g., vacuum). In such cases the radiative heat transfer between all surfaces depends only on the absolute temperature of the surfaces and the geometry. TPV cavities containing non-participating media with diffuse surfaces have been modelled by the net radiation method using view factors and a set of algebraic equations to calculate the radiative heat transfer between surfaces [6–11, 17–19]. The surfaces are usually subdivided until they can be approximated as having a uniform temperature. The net radiation method also allows the definition of spectrally selective surfaces. Enclosures with partially specular boundary conditions can be treated using specular view factors [9]. For a detailed analysis using view factors in planar and cylindrical (or tubular) geometries, the reader is referred to a chapter by Chubb [20].

Ray tracing for a TPV cavity: Generally the treatment of spectrally dependent and specularly reflective surfaces is complex, so that numerical models are typically utilised [9]. For cavities with specular surfaces, ray-tracing methods have been applied for TPV modelling using statistical methods. These methods are typically, but not exclusively based on Monte Carlo techniques [8, 21–27].

5.4.5 Radiative Heat Transfer with Participating Media

For semitransparent materials at high temperatures, unlike the more familiar opaque materials, the emission and absorption of radiation are bulk, rather than surface, phenomena [28]. This mechanism takes place in heat shield in the cavity (e.g., made of quartz glass). Equation 5.19 gives the spontaneous spectrally dependent emission of radiation in all directions by an isothermal volume element in a semitransparent medium, where α is the absorption coefficient and n the refractive index of the element [9].

$$n^2 \cdot \alpha \cdot 4\pi \cdot i_{b\lambda(\lambda,T_s)} \mathrm{d}V \mathrm{d}\lambda \tag{5.19}$$

Bulk radiation is then absorbed on its way towards the surface, some are internally reflected at this surface, and some refracted across the surface [28]. For one 1D radiative heat transfer the absorption of radiation is described by the Lambert-Beer law Eq. 5.7. Radiation within the semitransparent medium with refractive index n_1 (e.g., glass) incident on a surface of a second medium with refractive index n_2 (e.g., air) is either internally reflected or refracted across this surface. For optically smooth surfaces, the refraction across the surface can be described by *Snell's law* (Eq. 5.20).

$$\frac{\sin \theta_2}{\sin \theta_1} = \frac{n_1}{n_2} \tag{5.20}$$

Fresnel's equations give the specular reflection for polarised ($\rho \perp$ and ρ_{II}) and non-polarised (ρ) radiation depending on the angle of incidence for optically smooth surfaces (Eq. 5.21) [10]. Inserting Snell's law in Fresnel's equations can reduce the unknowns (n_1, n_2, θ_1 or θ_2) by one.

Table 5.3 Models for the radiative transfer equation, adapted from [10, 11]

Method	Angular resolution	Spatial resolution	Spectral resolution
Flux methods			
Multi flux approaches	Acceptable	Very good	Very good
Discrete transfer method	Acceptable	Very good	Very good
Discrete Ordinates (Sn)	Good	Very good	Very good
Moment methods			
Moment method	Not good	Very good	Very good
Spherical Harmonics (Pn)	Acceptable	(Very) Good	Very good
Zone methods	Acceptable	Good	Acceptable
Monte Carlo techniques	Very good	Good	Good
Numerical approaches			
Finite difference technique	Acceptable	Very good	Good
Finite element technique	Acceptable	Very good	Good

$$\rho_{II} = \left(\frac{n_1 \cos\theta_2 - n_2 \cos\theta_1}{n_1 \cos\theta_1 + n_2 \cos\theta_1}\right)^2 \rho_{\perp} = \left(\frac{n_1 \cos\theta_1 - n_2 \cos\theta_2}{n_1 \cos\theta_1 + n_2 \cos\theta_2}\right)^2$$
$$\rho = \frac{1}{2}\rho_{II} + \frac{1}{2}\rho_{\perp} \tag{5.21}$$

The radiative heat transfer within the glass can be modelled using the *Radiative Transfer Equation* (RTE) assuming no scattering (Eq. 5.22) [9–12]. The RTE is based on the conservation of radiant energy along a direction $\vec{s}$ at position $\vec{r}$ in an infinitesimally absorbing and emitting volume element [10, 11]. Physically, the first term on the right-hand side of Eq. 5.22 accounts for emission of radiation per unit volume. The second term accounts for the absorption of radiation per unit volume [12].

$$\nabla \cdot \left(I_{(\lambda,\vec{r},\vec{s})}\vec{s}\right) = \alpha_{(\lambda)}\left(n^2 i_{b\lambda(\lambda,T_s)} - I_{(\lambda,\vec{r},\vec{s})}\right) \tag{5.22}$$

Exact analytical solutions of this equation can only be found for idealised situations. For most other situations approximate solution methods are required. Howell and Mengüç compared these for multidimensional complex problems and regarded the Monte Carlo and Discrete Ordinates methods as having the best and a similar overall performance [11]. The four major approximation methods currently used are given by Modest and they are marked grey in Table 5.3 [10].

5.4.6 Radiative Heat Transfer Enhanced by the Refractive Index

It is well known that total blackbody emission into a medium with higher refractive index than $n = 1$ (other than vacuum or in good approximation air) scales with the refractive index squared, so that the total hemispherical emission becomes $n^2\sigma T^4$ according to the Stefan-Boltzmann law (Eq. 5.16) [10–12]. It has also been shown that the radiative heat transfer between two infinite opaque blackbody boundaries comprising a dielectric non-scattering and non-absorbing medium with refractive index n would be $n^2\sigma\ (T_h^4 - T_c^4)$ [10]. Examples, where the enhanced radiative heat transfer in dielectric media is important, are glass tank furnaces [28] and secondary solar concentrators using total internal reflection in a dielectric material [29].

The enhancement is also of interest for TPV conversion because a high radiative heat transfer translates into high electrical power densities. The enhancement was demonstrated experimentally. In an experiment a PV cell was illuminated by a light emitting diode (LED), where the gap in between was filled with either air ($n = 1$) or oil ($n = 1.5$) and the radiative heat transfer increased by about $n^2 = 2.25$ due to the use of oil [30]. It was concluded that the photon flux is limited by the lowest refractive index in the photonic cavity and scales with the minimum refractive index squared [30, 31]. This enhancement was termed *dielectric photon concentration* in [30]. Two approaches were considered, these are the dielectric insulator concept and the closed space or near-field concept (NF-TPV). Most research has concentrated on the latter [31]. Sect. 6.5.2 discusses details of the dielectric photon concentration concepts.

5.5 Combined Heat Transfer Modes

As pointed out in the previous sections, the three modes of heat transfer can occur in different areas of a TPV system. In some regions the heat transfer is not only by a single mode but also by combined modes. For example, in a gas filled cavity there can be conduction, convection and radiation heat transfer at the same time. The consideration of all modes makes modelling extremely complex. In order to reduce this complexity, simplifications need to be made. For example, less important modes can be neglected and purely the dominating modes can be considered. Combined modes require the coupling of different heat transfer equations. For example, coupled radiation and conduction introduces nonlinearity into the mathematical formulation. Under certain conditions these complex equations can be decoupled [8–10].

The remainder of this section focuses on combined *radiative and conductive heat transfer in semitransparent dielectric media* without convection. Bulk, rather than surface, radiation occurs in semitransparent dielectric media in the cavity. Often a temperature gradient forms across the dielectric media and this leads to conduction heat transfer. Hence, in these cases both modes conduction and radiation need to be considered. Examples are thick glass heat shield in the cavity [32], dielectric radiation guides between radiator and PV cell [33–35] and spectrally selective film radiators [36].

For greybody media with combined radiative and conductive heat transfer, there are two boundary cases: optically thin and optically thick. Optically thin is defined as $\alpha \cdot S \ll 1$ and optically thick as $\alpha \cdot S \gg 1$ assuming no scattering and a constant absorption coefficient α along a path length S [9]. In an enclosure with *optically thin* media, radiative heat transfer is dominated by surface-to-surface exchange [12]. In this case, radiative and conductive heat transfer can be decoupled, so that the total heat transfer between the surfaces is the sum of the two [8, 9]. For the *optically thick* approximation, radiative heat transfer within the medium can be treated as a localised phenomenon, little affected by surface boundaries, so that radiation emitted by a volume element is absorbed within another volume element before it can reach the boundaries. In this case, the diffusion (or Rosseland) approximation can be derived and has the same form as the Fourier conduction law. This enables a radiative conductivity to be defined that has the same format as the (molecular) conductivity with the unit W/(mK) (Eq. 5.23). The Rosseland approximation needs to be used with care near the boundaries [9, 10]. The effective conductivity is then the sum of both conductivities [8, 10, 28]. It becomes clear that coupling of conduction and radiation occurs purely for an intermediate optical thickness $\alpha \cdot$ S [12].

$$k_R = \frac{16n^2\sigma T^3}{3\alpha} \tag{5.23}$$

5.6 Summary

This chapter reviewed the basic theory about the three heat transfer modes: conduction, convection and radiation. The theory has been related to typical heat transfer phenomena in TPV systems. A key issue is the radiative heat transfer modelling of the cavity with diffuse and specular surfaces and spectrally dependent components such as radiators, filters and PV cells. The modelling complexity increases further, if combined heat transfer modes are considered for more realistic models of the system. The discussion about optical control in the next chapter focuses mainly on the radiative heat transfer in the cavity in terms of the spectral, angular and spatial distributions.

References

1. Guyer EC, Brownell DL (1999) Handbook of applied thermal design. Taylor & Francis, London
2. Baehr HD, Stephan K (2006) Wärme- und Stoffübertragung (in German), 5th edn. Springer, Berlin
3. Martinelli G, Stefancich M (2007) Solar cell cooling. In: Luque A, Andreev V (eds) Concentrator photovoltaics, Chap 7. Springer, Berlin, Heidelberg, pp 133–149
4. White DC, Hottel HC (1995) Important factors in determining the efficiency of TPV systems. Proceedings of the 1st NREL Conference on thermophotovoltaic generation of electricity. Copper Mountain, Colorado, 24–28 July 1994. American Institute of Physics, pp 425–454

5. Coutts TJ (1999) A review of progress in thermophotovoltaic generation of electricity. Renew Sustain Energy Rev 3(2–3):77–184
6. Aschaber J, Hebling C, Luther J (2001) Modelling of a thermophotovoltaic system incuding radiation and conduction heat transfer. Proceedings of the 17th European photovoltaic solar energy conference, pp 186–189
7. Aschaber J, Hebling C, Luther J (2002) The challenge of realistic tpv system modelling. Proceedings of the 5th Conference on thermophotovoltaic generation of electricity, Rome, Italy, 16–19 Sept 2002. American Institute of Physics, pp 79–90
8. Aschaber J, Hebling C, Luther J (2003) Realistic modelling of TPV systems. Semicond Sci Technol 18:158–164
9. Siegel R, Howell J (2001) Thermal radiation heat transfer, 4th edn. Taylor & Francis, New York, London
10. Modest MM (1993) Radiative heat transfer. McGraw-Hill, New York
11. Howell JR, Mengüç MP (1998) Radiation. In: Rohsenow WM et al. (eds) Handbook of heat transfer, Chap 7, 3rd edn. McGraw-Hill, New York, pp 7.1–7.100
12. Viskanta R, Anderson EE (1975) Heat transfer in semitransparent solids. Advances in Heat Transfer 11:317–441
13. Biter PJ, Georg KA, Phillips JE (1997) A TPV system using a gold filter with $CuInSe_2$ solar cells, 3rd NREL conference on thermophotovoltaic generation of electricity, Denver, Colorado, 18-21 May 1997. American Institute of Physics, pp 443–459
14. Les J, Borne T, Cross D, Gang Du, Edwards DA, Haus J, King J, Lacey A, Monk P, Please C, Hoa T (2000) Interference Filters for Thermophotovoltaic Applications. In Proceedings of the 15th workshop on mathematical problems in industry, University of Delaware, US, June 1999
15. Burger DR (1997) Modeling the TPV system optical cavity, Proceedings of the 3rd NREL conference on thermophotovoltaic generation of electricity, Denver, Colorado, 18–21 May 1997, American Institute of Physics, pp 535–546
16. Schroeder KL, Rose MF, Burkhalter JE (1997) An improved model for TPV performance predictions and optimization, Proceedings of the 3rd NREL conference on thermophotovoltaic generation of electricity, Denver, Colorado, 18–21 May 1997. American Institute of Physics, pp 505–519
17. Bitnar B, Durisch W, Mayor J-C, Sigg H, Tschudi HR, Palfinger G, Gobrecht J (2003) Record electricity-to-gas power efficiency of a silicon solar cell based TPV system, Proceedings of the 5th conference on thermophotovoltaic generation of electricity, Rome, Italy, 16–19 Sept 2002. American Institute of Physics, pp 18–28
18. Bitnar B, Durisch W, Mayor J-C, Sigg H, Tschudi HR, Palfinger G, Gobrecht J (2001) Development of a small TPV prototype system with an efficiency more than 2%, Proceedings of the 17th European photovoltaic solar energy conference, Munich, 22–26 Oct 2001. WIP
19. Good BS, Chubb DL (1997) Effects of geometry on the efficiency of TPV energy conversion, Proceedings of the 3rd NREL conference on thermophotovoltaic generation of electricity, Denver, Colorado, 18–21 May 1997. American Institute of Physics, pp 487–503
20. Chubb D (2007) Fundamentals of Thermophotovoltaic Energy Conversion. Elsevier Science, Amsterdam
21. Fraas L, Avery J, Malfa E, Wuenning JG, Kovacik G, Astle C (2003) Thermophotovoltaics for combined heat and power using low NOx gas fired radiant tube burners. Proceedings of the 5th conference on thermophotovoltaic generation of electricity, Rome, 16–19 Sept 2002. American Institute of Physics, pp 61–70
22. Lindberg E (2002) TPV Optics Studies—On the Use of non-imaging optics for improvement of edge filter performance in thermophotovoltaic applications. Doctoral thesis, Swedish University of Agricultural Science, Uppsala
23. Gethers CK, Ballinger CT, Postlethwait MA, DePoy DM, Baldasaro PF (1997) TPV efficiency predictions and measurements for a closed cavity geometry, Proceedings of the 3rd NREL conference on thermophotovoltaic generation of electricity. Denver, Colorado, 18–21 May 1997. American Institute of Physics, pp 471–486

24. Ballinger CT, Charache GW, Murray CS (1999) Monte Carlo analysis of a monolithic interconnected module with a back surface reflector, Proceedings of the 4th NREL conference on thermophotovoltaic generation of electricity. Denver, Colorado, 11–14 Oct 1998. American Institute of Physics, pp 161–174
25. Thomas RM, Wernsman BR (2001) Thermophotovoltaic devices and photonics modeling. Optics and Photonics News 12(8):40–44
26. Wernsman B, Mahorter RG, Thomas RM (2002) Optical cavity effects on TPV efficiency. Proceedings of the 5th conference on thermophotovoltaic generation of electricity, Rome, 16–19 Sept 2002. American Institute of Physics, pp 277–286
27. Gopinath A, Aschaber J, Hebling C, Luther J (2004) Modeling of radiative energy transfer and conversion in a TPV power system, Proceedings of the 6th international conference on thermophotovoltaic generation of electricity, Freiburg, Germany, 14–16 June 2004. American Institute of Physics, pp 162–168
28. Gardon R (1961) A review of radiant heat transfer in glass. J Am Ceram Soc 44:305–311
29. Winston R, Cooke D, Gleckman P, Krebs H, OGallagher J, Sagie D (1990) Sunlight brighter than the sun. Nature 346:802
30. Baldasaro PF, Fourspring PM (2003) Improved thermophotovoltaic (TPV) performance using dielectric photon concentrations (DPC). Lockheed Martin Inc., US, Technical Report, LM-02K136
31. DiMatteo R, Greiff P, Seltzer D, Meulenberg D, Brown E, Carlen E, Kaiser K, Finberg S, Nguyen H, Azarkevich J, Baldasaro P, Beausang J, Danielson L, Dashiell M, DePoy D, Ehsani H, Topper W, Rahner K, Siergiej R (2004) Micron-gap ThermoPhotoVoltaics (MTPV), Proceedings of the 6th international conference on thermophotovoltaic generation of electricity, Freiburg, Germany 14–16 June 2004. American Institute of Physics, pp 42–51
32. Bauer T, Forbes I, Penlington R, Pearsall N (2005) Heat transfer modelling in thermophotovoltaic cavities using glass media. Sol Energy Mater Sol Cells 88(3):257–268
33. Goldstein MK, DeShazer LG, Kushch AS, Skinner SM (1997) Superemissive light pipe for TPV applications, Proceedings of the 3rd NREL conference on thermophotovoltaic generation of electricity, Denver, Colorado, 18–21 May 1997. American Institute of Physics, pp 315–326
34. DeShazer LG, Kushch AS, Chen KC (2001) Hot dielectrics as light sources for TPV devices and lasers, NASA Tech Briefs Magazine, [Online] Available at: http://www.nasatech.com/. Accessed 10 Sept 2004
35. Goldstein MK (1996) Superemissive light pipes and photovoltaic systems including same, Quantum Group Inc., US Patent 5500054
36. Chubb DL, Good BS, Clark EB, Zheng C (1997) Effect of temperature gradient on thick film selective emitter emittance, Proceedings of the 3rd NREL conference on thermophotovoltaic generation of electricity, Denver, Colorado, 18–21 May 1997. American Institute of Physics, pp 293–313

Chapter 6
Cavity Design and Optical Control

6.1 Introduction

The key role of an effective system is the thermal design of the cavity formed by the radiator, mirrors and PV cells, as well as a good match of the these components. The important TPV figures of merit, high efficiency and high electrical power density critically depend on the heat transfer between the radiator and the PV cell in the TPV cavity (see Fig. 5.1). Radiative heat transfer in the cavity needs to be optimised in terms of the spectral, angular and spatial radiation distributions.

For optimum PV cell performance, the cells need *spatially uniform* illumination. Typically the cells are series connected to increase the voltage. A lower current from one cell, due to spatially non-uniform illumination, reduces the power output from the whole string.

The *angle* of the radiation in the TPV cavity is also of significance. Flat surfaces of dielectric materials increasingly reflect radiation with incident angles away from the normal (or large zenith angles) according to Fresnel's equations (see Sect. 5.4.5). This can result in lower radiative heat transfer rates from the radiator to the PV cell. An example would be a PV cell facing a blackbody radiator, where from the definition, radiation leaves the blackbody radiator at all angles (Lambert's cosine-law). However, radiation incident at the PV cell surface with large zenith angles is increasingly reflected (assuming a flat PV cell surface without light trapping). This leads to a lower radiative heat transfer compared to two blackbody surfaces facing each other. Similarly, radiation is reflected by heat shield or filters placed in the cavity. Collimators can align diffuse radiation so that radiation incident at surfaces can have smaller zenith angles. As a result, TPV systems may be designed with higher radiative heat transfer rates. In addition to the angle dependent surface reflection, the radiative surface properties (e.g., emissivity and reflectivity) of real components (e.g., radiator, shield, filters and PV cell) usually have both angular *and* spectral dependences. For example, dielectric filters become spectrally less selective for angles away from normal. This dual dependency adds to the complexity of spectral control in the TPV cavity.

T. Bauer, *Thermophotovoltaics*, Green Energy and Technology,
DOI: 10.1007/978-3-642-19965-3_6,

The *spectrum* of the absorbed radiation by the PV cell has a major impact on the cell conversion efficiency. Most of the remaining discussion in this chapter focuses on this aspect using the term spectral control. In the review and discussion about PV cells (Sect. 4.2), the cell efficiency was split into partial efficiencies. The PV cell related efficiencies, namely the voltage factor, collection efficiency, fill factor and PV cell array efficiency η_{Array}, need to be optimised for a specific PV cell type and system design. In such optimised systems, the spectral match of the absorbed power by the PV cell to its bandgap is of interest and this spectral control is described by the ultimate efficiency discussed and modelled in the next section.

6.2 Ultimate Efficiency and Power Density (Upper Limits)

Several models have been developed to predict an upper limit of the efficiency and power density, as well as to identify the ideal PV cell bandgap. There are large differences in the assumptions of these models and, as can be expected, also in the modelling result. Coutts gives an overview of different models [1, 2] and Cody compares efficiencies for different models without spectral control [3]. Some models are based on empirical values for the reverse saturation current density including work by Woolf [4, 5], Wanlass et al. [6, 7], Caruso and Piro [8] and Iles et al. [9]. Some modelling refers specifically to solar TPV conversion, such as work by Höfler et al. [10–12] and others [13–16]. De Vos modelled the solar PV conversion as an endoreversible heat engine [17–19]. This model has been adapted to TPV conversion by Gray and El-Husseini [20] and Heinzel et al. [21, 22]. In general, there is limited work on the combined effect of spectral suppression of radiation above and below the PV cell bandgap. Usual assumptions include no above and no sub-bandgap suppression (full spectrum case) [3, 15, 23], only suppression of sub-bandgap radiation [15, 21–24], perfect sub-bandgap and variable above-bandgap suppression [13] and specific percentage or band assumptions about the sub and above-bandgap suppression [1, 9, 20]. The listed literature shows the diversity of the modelling assumptions, without the aim to be exhaustive. Hence, one major difficulty in modelling TPV efficiency and electrical power density comprehensively is the number of parameters. Major parameters include the radiator temperature T_s, the PV cell bandgap $h\nu_g$, as well as the suppression of sub and above-bandgap radiation.

Some authors used the substitution $x = h\nu/kT_s$ (or $x_g = h\nu_g/kT_s$) [2, 24–26]. This substitution can reduce the parameters by one and it is also used within the following text. The focus is set on the simultaneous above and sub-bandgap suppression on an ultimate efficiency level. This loss mechanism corresponds to the free and hot carrier heating in the PV cell . The approach allows insights into spectral control in general and can give an upper limit for the efficiency and electrical power density. The major simplifying assumptions made in this modelling are: a single bandgap PV cell, illumination by a blackbody, no reflection losses, a view factor of unity and all other partial efficiencies are unity (η_{OC}, η_{QE}, η_{FF}, η_{Array} and η_{Cavity}, Sect. 4.2). In the following subsections, first, the ultimate efficiency of the solar PV conversion process as presented by Shockley and Queisser is repeated [27]. Subsequently, the modelling is extended to the more general TPV case.

6.2.1 Solar PV Conversion

The sun can be approximately modelled as a blackbody radiator with a temperature $T_s = 5800$ K. Equation 6.1 gives the number of photons with energies greater than $h\nu_g$ leaving this blackbody per unit area and per unit time [27]. The number of photons is obtained by the division of the radiation intensity with the photon energy $h\nu$.

$$Q_{s(\nu_g,T_s)} = \frac{2\pi}{c_0^2}\int\limits_{\nu_g}^{\infty}\frac{\nu^2}{e^{\frac{h\nu}{kT_s}}-1}\mathrm{d}\nu = \frac{2\pi}{c_0^2}\cdot\left(\frac{kT_s}{h}\right)^3\int\limits_{x_g}^{\infty}\frac{x^2}{e^x-1}\mathrm{d}x \tag{6.1}$$

Integration of Planck's radiation law yields the Stefan-Boltzmann law given in Eq. 6.2 (see also Sect. 5.4.3, Eqs. 5.15 and 5.16) [27, 28].

$$\begin{aligned} I_{b(T_s)} &= \frac{2\pi h}{c_0^2}\int\limits_0^{\infty}\frac{\nu^3}{e^{\frac{h\nu}{kT_s}}-1}\mathrm{d}\nu \\ &= \frac{2\pi h}{c_0^2}\cdot\left(\frac{kT_s}{h}\right)^4\int\limits_0^{\infty}\frac{x^3}{e^x-1}\mathrm{d}x = \frac{2\pi h}{c_0^2}\cdot\left(\frac{kT_s}{h}\right)^4\frac{\pi^4}{15} \\ &= \frac{2\pi^5k^4}{15h^3c_0^2}T_s^4 = \sigma T_s^4 \end{aligned} \tag{6.2}$$

Assuming the sun as a hot sphere with radius r_s, the values of Q_s and I_b drop by a factor of r_s^2/r_{se}^2 at sun-earth distance r_{se}. Equation 6.3 gives the PV power density. One may calculate the solar radiation density at the outer earth atmosphere as the product of σT_s^4 and r_s^2/r_{se}^2 equal to about 1380 W/m^2. Equation 6.4 defines the ultimate PV cell efficiency at the outer earth atmosphere, assuming each photon with the energy $h\nu$ greater than $h\nu_g$ contributes to the electricity output $h\nu_g$. The ultimate solar efficiency $\eta_{UE,solar}$ has a maximum value of about 44% at $x_g \approx 2.2$ [27]. For this simplified model, one may also calculate the ideal bandgap for solar PV conversion as $h\nu_g = 1.1$ eV using the definition of the substitution $x_g = h\nu_g/kT_s$. It should be considered that realistic terrestrial solar spectra with absorption bands result in other ideal bandgap values.

$$P_{solar} = \frac{r_s^2}{r_{se}^2}\cdot h\nu_g\cdot Q_{s(\nu_g,T_s)} \tag{6.3}$$

$$\eta_{UE,solar} = \frac{h\nu_g\cdot Q_{s(\nu_g,T_s)}}{I_{b(T_s)}} = \frac{x_g\int_{x_g}^{\infty}\frac{x^2}{e^x-1}\mathrm{d}x}{\int_0^{\infty}\frac{x^3}{e^x-1}\mathrm{d}x} \tag{6.4}$$

6.2.2 TPV Conversion Without Spectral Control

For TPV conversion, the radiator and the PV cell can be closely arranged and ideally no radiation is lost, so that the factor r_s^2/r_{se}^2 in Eq. 6.3 can be assumed unity (view factor is unity). On the other hand, Q_s values are smaller for TPV compared to non-concentrator solar PV conversion because of the lower radiator temperatures for TPV conversion. Overall, higher power densities are usually obtained for TPV conversion, in the order of W/cm^2, compared to non-concentrator solar PV conversion in the order of 0.01 W/cm^2.

It has been pointed out that the ultimate solar efficiency η_{solar} Eq. 6.4 is also valid for TPV conversion with the assumption of *no* spectral control [1, 29]. It becomes clear that the ideal PV cell bandgap energies for TPV are smaller compared to solar PV conversion, since the substitution $x_g = h\nu_g/kT_s$ becomes maximum at the constant value of about 2.2 and the radiator temperatures are lower for TPV conversion. Using T_s values in the range from 1300 to 2000 K results in $h\nu_g$ values from 0.25 to 0.38 eV. Currently, it is challenging to obtain cell related efficiencies $\eta_{OC}\ \eta_{FF} > 0$ from such low bandgap cells (Fig. 4.5). For this reason, it is currently not possible to develop efficient TPV systems without spectral control.

6.2.3 TPV Conversion with Spectral Control

Unlike solar PV conversion, TPV conversion offers the possibility to suppress or recover photons using spectral control methods (e.g., filters and selective radiators). It is assumed here that all photons absorbed by the PV cell be within a spectral band, with some blackbody radiation below and some above the PV bandgap $h\nu_g$. Figure 6.1 shows an example of this band model using the normalised Planck's radiation law Eq. 6.5 and the substitution $x = h\nu/kT_s$, where $x-$ is the lower and $x+$ the upper band limit. The normalised bandgap x_g is assumed to be in between these two boundaries.

$$I_{b\nu(\nu,T_s)} = \frac{2\pi h}{c_0^2}\frac{\nu^3}{e^{\frac{h\nu}{kT_s}}-1} = \frac{2\pi h}{c_0^2}\left(\frac{kT_s}{h}\right)^3\frac{x^3}{e^x-1} \tag{6.5}$$

It should be considered that a change in radiator temperature T_s in Eq. 6.5 does not change the shape of the plotted function in Fig. 6.1, but scales the function equally for all values of x (Eq. 6.5). Practically, the absorbed radiation by the PV cell will differ from the band model. Nevertheless, the band approach can be justified since for other spectra an equivalent band model can be calculated, which generates the same amount of heat and electricity [31].

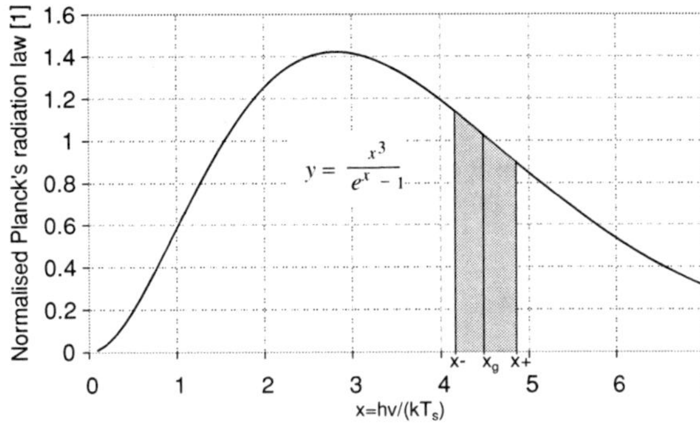

Fig. 6.1 Plot of the normalised Planck's radiation law [30]. The substitution $x = h\nu/(kT_s)$ was used. An example of the band model is shown

Similar to the solar PV case (Eq. 6.1), Q_T can be defined as the number of photons with energies greater than $h\nu_g$ and below the band limit $\nu+$ (Eq. 6.6). The radiation absorbed by the PV cell I_T can be defined for the TPV case using the spectral band model (Eq. 6.7). In the following text, it will be shown that splitting the integral at x_g is a convenient mathematical step (Eq. 6.7).

$$Q_{T(\nu_g,T_s,\nu+)} = \frac{2\pi}{c_0^2}\int\limits_{\nu_g}^{\nu+}\frac{\nu^2}{e^{\frac{h\nu}{kT_s}}-1}\mathrm{d}\nu = \frac{2\pi}{c_0^2}\cdot\left(\frac{kT_s}{h}\right)^3\int\limits_{x_g}^{x+}\frac{x^2}{e^x-1}\mathrm{d}x \tag{6.6}$$

$$\begin{aligned} I_{T(T_s,\nu_+,\nu-)} &= \frac{2\pi h}{c_0^2}\int\limits_{\nu-}^{\nu+}\frac{\nu^3}{e^{\frac{h\nu}{kT_s}}-1}\mathrm{d}\nu = \frac{2\pi h}{c_0^2}\cdot\left(\frac{kT_s}{h}\right)^4\int\limits_{x-}^{x+}\frac{x^3}{e^x-1}\mathrm{d}x \\ &= \frac{2\pi h}{c_0^2}\cdot\left(\frac{kT_s}{h}\right)^4\left[\int\limits_{x-}^{x_g}\frac{x^3}{e^x-1}\mathrm{d}x + \int\limits_{x_g}^{x+}\frac{x^3}{e^x-1}\mathrm{d}x\right] \end{aligned} \tag{6.7}$$

The spectral band limits $x+$ and $x-$ need to be viewed together with the normalised bandgap x_g in order to be interpreted. As a single value they are less meaningful. A more meaningful single value is the ratio of radiation within the spectral band below the bandgap divided by the total radiation below the bandgap ($r-$). For the above-bandgap radiation, similarly $r+$ is defined (Eq. 6.8). These newly defined ratios have a range from 0 to 1.

$$\begin{aligned} r- &= \frac{\int_{x-}^{x_g}\frac{x^3}{e^x-1}\mathrm{d}x}{\int_0^{x_g}\frac{x^3}{e^x-1}\mathrm{d}x} \\ r+ &= \frac{\int_{x_g}^{x+}\frac{x^3}{e^x-1}\mathrm{d}x}{\int_{x_g}^{\infty}\frac{x^3}{e^x-1}\mathrm{d}x} \end{aligned} \tag{6.8}$$

The relation between $r+$ and $x+$ depending on x_g can then be computed numerically from (Eq. 6.8). For example, if x_g and $r+$ is known, the unknown $x+$ can be computed. With this consideration the TPV power density and efficiency can be defined (Eqs. 6.9 and 6.10). Computation of values between $r-$ and $x-$ is not necessary, since the definition of $r+$ and $r-$ (Eq. 6.8) can be inserted into the definition of I_T (Eq. 6.7) by splitting the integral at x_g.

$$P_{TPV} = h\nu_g \cdot Q_{T(T_s,x_g,x+)} = \frac{2\pi h}{c_0^2}\left(\frac{kT_s}{h}\right)^4 x_g \int_{x_g}^{x+} \frac{x^2}{e^x - 1}dx \tag{6.9}$$

$$\begin{aligned}\eta_{UE,TPV(x_g,x-,x+)} &= \frac{h\nu_g \cdot Q_T}{I_T} = \frac{x_g \int_{x_g}^{x+} \frac{x^2}{e^x-1}dx}{\int_{x-}^{x+} \frac{x^3}{e^x-1}dx} \\ &= \frac{x_g \int_{x_g}^{x+} \frac{x^2}{e^x-1}dx}{r- \int_0^{x_g} \frac{x^3}{e^x-1}dx + r+ \int_{x_g}^{\infty} \frac{x^3}{e^x-1}dx}\end{aligned} \tag{6.10}$$

In the following, computed results of the (Eqs. 6.9) and 6.10 are presented. The x-axis has been always used to plot the normalised bandgap $x_g = h\nu_g/kT_s$ and the parameters $r-$ and $r+$ are altered parametrically.

The *power density* does not depend on the sub-bandgap suppression (Eq. 6.9) because photons within energies below the bandgap cannot be converted. Hence, the ideal normalised bandgap for the maximum electrical power density is $x_g = 2.2$, if no above-bandgap radiation is suppressed. This is the same value as for the ultimate efficiency for solar PV conversion. Equations 6.4 and 6.9 give the same maximum for x_g, because the denominator integral of Eq. 6.4 is simply the constant $\pi^4/15$ [28]. Figure 6.2 shows the decrease of the electrical power density with increasing above-bandgap suppression.

The ideal normalised bandgap x_g, or the maximum of the power density function, shifts to even smaller values, if radiation above the bandgap is suppressed (Fig. 6.2). As discussed, the value $x_g = 2.2$ already results in very low ideal PV cell bandgap energies. For this reason, it can be concluded that above-bandgap suppression is not desirable from the viewpoint of electrical power density. Bearing in mind that typical GaSb cell and radiator temperature combinations result in x_g values from 4.8 to 5.4 [32–35] and Si cell based systems from 5.4 to 7.5 [36–39], there are two ways to increase the power density. These are cells with reduced bandgaps and higher radiator temperatures. Figure 6.2 shows that smaller x_g values (or smaller bandgaps $h\nu_g$) that are closer to the maximum at around 2.2 could increase the power density largely. Equation 6.9 shows the strong increase of the power density with the fourth power of the absolute temperature.

The *efficiency* depends on all three modelled parameters, the normalised PV cell bandgap x_g (1), the above-bandgap suppression $r+$ (2) and the sub-bandgap suppression $r-$ (3). The three plots in Fig. 6.3 vary in terms of the above-bandgap suppression among each other. Other assumptions remain the same between these plots. Several conclusions can be drawn. Both above and sub-bandgap suppression

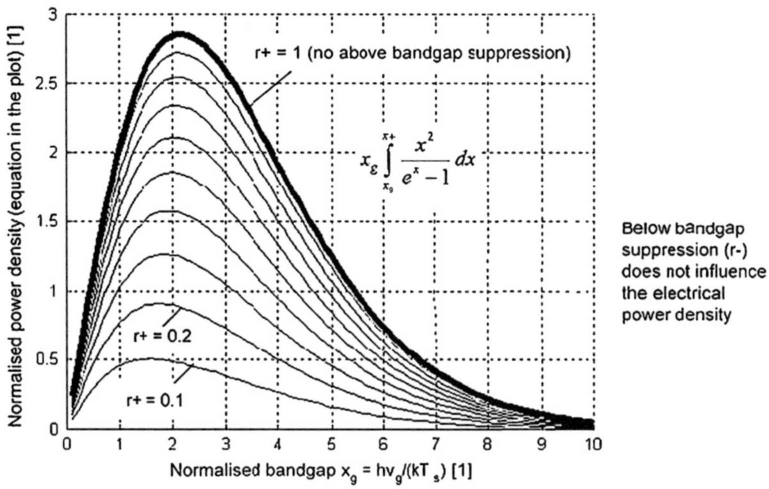

Fig. 6.2 Plot of the normalised power density as a function of *xg* and r+ [30]. The bold line shows the normalised power density without above-bandgap suppression ($r+ = 1$). The above-bandgap suppression varies parametrically in steps of 0.1 from $r+ = 0.1$ to $r+ = 1$

can increase the maximum efficiency. The theoretical case of monochromatic illumination with the bandgap energy of the PV cell may be considered ($r+ = 0$, $r- = 0$). In such a case the ultimate efficiency would become 100% at an electrical power density of zero (not shown). Hence, full above-bandgap suppression is unrealistic. Also full sub-bandgap suppression ($r- = 0$) leads to unrealistic results. Here, the efficiency function increases monotonically with the normalised bandgap x_g (Fig. 6.3) [24]. Woolf also highlighted in an early work the impact of very small parasitic sub-bandgap absorption on the TPV efficiency [4]. The importance of sub-bandgap suppression (or spectral control) can be strengthened using a typical example. A TPV system using a GaSb cell ($h\nu_g = 0.72$ eV) and a radiator temperature of 1500 K results in $x_g = 5.57$. Assuming no above and no sub-bandgap suppression the ultimate efficiency is about 15% (Fig. 6.3 top, solar PV case). Assuming that only 10% of the sub-bandgap radiation is absorbed by the PV cell (or 90% of the sub-bandgap radiation is suppressed), the ultimate efficiency increases considerably to a value of about 55% (Fig. 6.3 top, $r- = 0.1$). Figure 6.3 also shows that there is scope for even higher ultimate efficiencies, if less than 10% sub-bandgap radiation is absorbed by the cell. Tandem cells , not discussed in detail here, offer an additional path to increase the ultimate efficiency further (see Sect. 4.7.1).

In the case of simultaneous above and sub-bandgap suppression (Fig. 6.3 bottom, $r- = 0.1$) the ultimate efficiency even decreases below the case with sub-bandgap suppression only (from 15% to 10%). This is the result of a shift of the efficiency function to smaller normalised ideal bandgaps for additional above-bandgap suppression. Generally, it can be also seen that the sub-bandgap suppression broadens and the above-bandgap suppression sharpens the efficiency

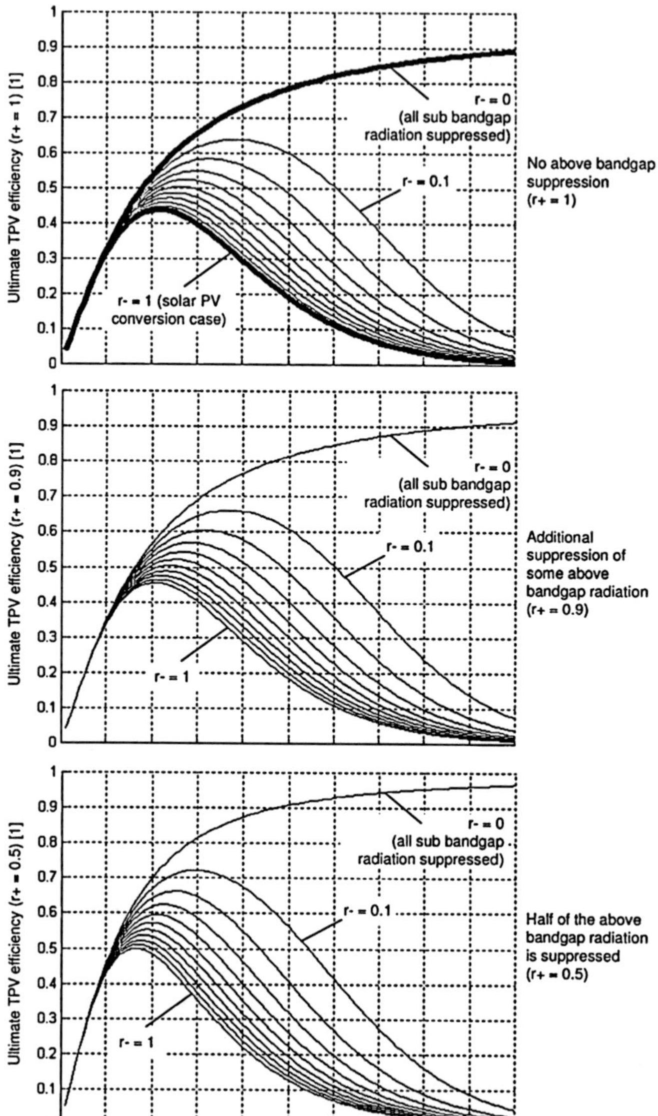

Fig. 6.3 Three plots of the ultimate efficiency as a function of x_g, $r-$ and $r+$ [30]. For each plot the sub-bandgap suppression varies parametrically in steps of 0.1 from $r- = 1$ (solar PV case shown bold for the first plot) to the $r- = 0$ (all radiation suppressed also shown bold for the first plot). Above-bandgap radiation has been additionally suppressed for the two plots in the *middle* and *bottom*

curves compared to the solar PV case ($r- = 1$, $r+ = 1$). A broader efficiency curve is generally desirable because this can result in high efficiencies even if x_g values are not matched to the maximum of the function as discussed for the GaSb example with $x_g = 5.57$.

6.2.4 Summary

Generally, the ultimate efficiency shows that complete sub or above-bandgap suppression ($r- = 0$, $r+ = 0$) lead to unrealistic results, so that $r-$ and $r+$ values larger than 0 are of interest. Both, efficiency and electrical power density could be increased for smaller PV cell bandgaps $h\nu_g$ or higher radiator temperatures T_s compared to the current situation with typical values of x_g larger than 5. Historically, there has been a trend to smaller bandgaps in TPV cell development and this trend may be expected to continue. However, the ultimate efficiency cannot be seen as isolated from the other partial efficiencies. For smaller PV cell bandgaps $h\nu_g$, the fill factor and voltage factor decrease significantly (see Fig. 4.5). Hence, for smaller bandgaps the trade-off between an increase of the ultimate efficiency and a decrease of other partial efficiencies needs to be assessed.

There is generally agreement that sub-bandgap suppression can enhance the ultimate efficiency considerably assuming currently utilised x_g values. The modelling presented in the previous subsection showed that the maximum of the efficiency function shifts to higher normalised bandgaps x_g and the function broadens.

It is obvious that simultaneous above and sub-bandgap suppression can enhance the efficiency further [1]. For example, PV cells have demonstrated very high efficiencies under spectrally narrow laser illumination that was matched to the cell bandgap. In spite of this, it can be concluded that above-bandgap suppression is in most cases not desirable for TPV conversion. The results of the modelling (Fig. 6.3) show that the efficiency cannot be enlarged very much using additional above-bandgap suppression. Furthermore, it needs to be considered that typical TPV systems have values x_g above 5. Typically, these systems do not operate in the maximum of the efficiency function because below bandgap suppression is usually not better than 90% (or $r- = 0.1$). Finally, as already mentioned, above bandgap suppression reduces the power density. As a result of the presented assessment of simultaneous sub and above-bandgap suppression, this text defines the figures of merit for the spectral control as follows:

- High share of in-band radiation absorbed by the PV cell; alternatively this can be termed: high share of short wavelength radiation, high share of photons with energies larger than the bandgap energy or low above-bandgap suppression.
- Low share of out-of-band radiation absorbed by the PV cell; alternatively this can be termed: low share of long wavelength radiation, low share of photons with energies lower than the bandgap energy or high sub-bandgap suppression.

With these assumptions, it needs to be considered that the maximum of the power density function remains the same (x_g is maximised at about 2.2), whereas the maximum of the efficiency function shifts to larger x_g values. Therefore, there is no common ideal PV cell bandgap to maximise both, the ultimate efficiency and the electrical power density simultaneously. Hence, a trade-off between high efficiency and high electrical power density occurs due to sub-bandgap suppression. In other words, a TPV system can be specifically designed to achieve a high power density or high efficiency. For example, industrial waste heat could be freely available and it would be sufficient to maximise the electricity output. The efficiency, and hence sub-bandgap suppression, would not be the driving factor. On the other hand, for example a portable power supply unit aims for a long operation time and this would require the efficiency to be maximised (e.g., by suppressing most sub-bandgap radiation). This situation is fundamentally different from solar PV conversion. For solar PV the power density and the efficiency are maximised simultaneously.

6.3 TPV Cavity Arrangements

The *location of the filter* in the cavity can be considered as critical. For close location near the radiator, overheating of filters has been reported [40, 41]. Hence, there has been a historical preference to place the filter first closer to the PV cell and finally on the front surface of the cell. Mounted on top, the filter can be cooled via the PV cell [42, 43]. On the other hand, heat shield designs made of quartz glass could generally withstand the temperatures in the TPV cavity. There have been several geometric configurations of TPV cavities and these are discussed in the following subsections.

6.3.1 Configuration with Minimum Mirror Utilisation

Figure 6.4 shows configurations where the radiator can be completely surrounded by the PV cells. A cuboid system using a radioisotope heat source has been examined. For such configuration it is theoretically feasible to cover all six faces with PV cells, although currently for some faces mirrors are considered [44, 45]. Also solar TPV systems may be designed with a cavity made of PV cells, a centred radiator and a small aperture to receive the radiation [46, 47]. The minimisation of the mirror utilisation has the advantage that radiation is necessarily transferred to the PV cell and heat losses, via parasitic absorption, in the mirror area is minimised. A drawback is the limitation of the power density, since the area ratio of radiator to PV cell is always smaller than one. In addition practical implementation should pose some engineering challenges, such as holding of the radiator, spherical system design or non-uniform PV cell illumination for a cube

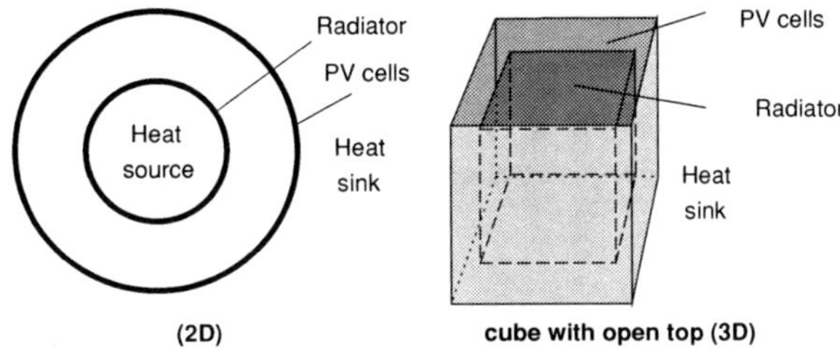

Fig. 6.4 Cavity configurations with minimum mirror utilisation

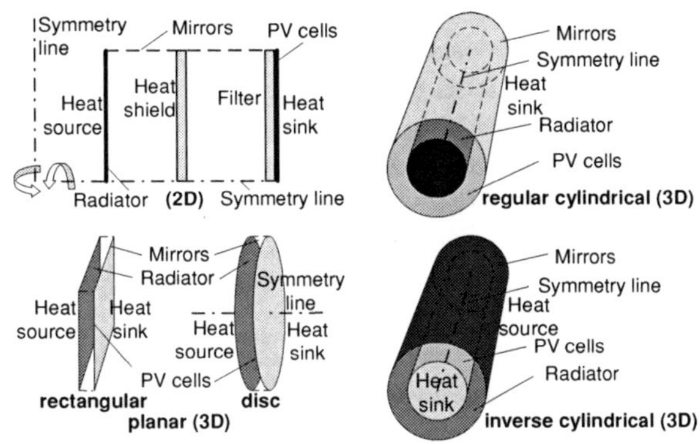

Fig. 6.5 Cylindrical and planar cavity configurations

design. Half spherical systems have been also proposed for solar TPV conversion [10, 48].

6.3.2 Tubular and Planar Configurations with Mirrors

By far the most common cavity arrangements are planar and cylindrical designs. These designs are a suitable compromise between convenient engineering solutions and acceptable small mirror areas.

Figure 6.5 on the top left shows a common two-dimensional generic schematic of these designs. The cavity comprises the radiator, the mirrors and the PV cell with a filter on the front surface. Often a quartz glass heat shield has been placed in between the radiator and the PV cells. The mirror area can be minimised by making the area of radiator and PV cell large compared to their distance. Mirrors can minimise cavity losses and direct the radiation to the PV cell. Engineering of the mirror area is usually challenging (see Sect. 6.4.2). The two-dimensional generic schematic represents some three-dimensional cavities, such as the planar disc and the cylindirical tubes, if different symmetry conditions are introduced (e.g., rotational symmetry). The *regular cylindrical* design, or also called concentric tubular design, has been most commonly chosen (Fig. 6.5 on the top right) [32, 49–60]. For combustion systems, it has the advantage that radiant tube burners forming the inner tube are readily available (see Sect. 8.4.1). The *inverse*

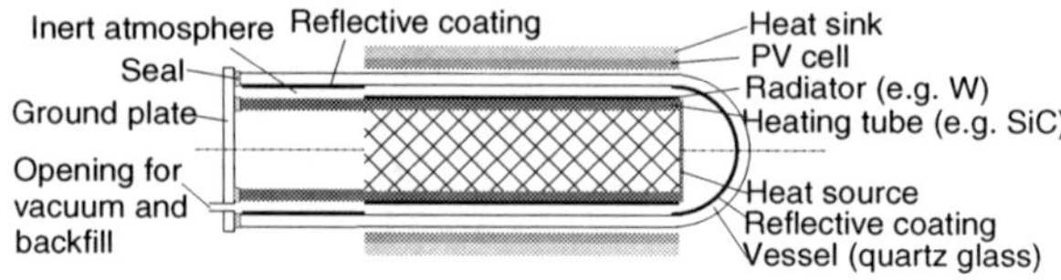

Fig. 6.6 Example of a cross-section of a regular cylindrical cavity configuration with inert atmosphere and a glass vessel

cylindrical configuration with the PV cells on the inner tube and the radiator forming the outer tube has received far less attention. It is attractive to place the inverse configuration in a (existing) high-temperature chamber, such as for industrial waste heat recovery and self-powered devices [61, 62]. Solar and combustion systems using disc and rectangular *planar* designs have been developed (Fig. 6.5 on the bottom left) [63–68] The fundamental *differences of concentric and planar* configurations in terms of radiative heat transfer aspects have been assessed. Aspects of interest included view factors, spatial radiation uniformities and cavity losses. With the exception of the power density, the regular cylindrical geometry has been considered more advantageous than the planar geometry [40, 69, 70].

For the regular cylindrical geometry, solar and combustion systems with an inert gas atmosphere (or vacuum) have been examined [71–73]. Figure 6.6 shows a possible engineering solution of such configuration. One advantage is that conduction and convection heat losses via the cavity gas can be minimised. In addition, suitable spectrally selective metal radiators, such as tungsten, cannot operate in an oxidizing environment, but only in an inert atmosphere at high temperatures. This configuration introduces some additional engineering aspects. The design includes a leak-proof vessel that contains the high-temperature radiator. Usually a seal or solder (e.g., graphite and glass-to-metal) that cannot operate at very high temperatures needs to be included. Engineering aspects of the seal or solder include the type, the location, the mismatches in thermal expansion and the cooling method. At the current stage, a heating tube made of silicon carbide seems to be a good choice for combustion systems. Conduction losses via this tube can also occur. Also, thermally stable mounting techniques for the selective radiator (e.g., tungsten foil) on this tube need to be identified. At both ends of the radiator, heat losses to the vessel (e.g., quartz glass) via radiation need to be minimised. Additional reflective coatings and radiation shields can reduce losses via this path. This discussion points out that several engineering aspects need to be addressed in order to successfully design an inert TPV cavity.

6.3.3 Collimator and Concentrator in the Cavity

As discussed in the last subsection the geometry of the cavity has an impact on the spatial and angular distribution of the radiation. Although there has been limited research in this area, specifically designed cavity geometries may improve the

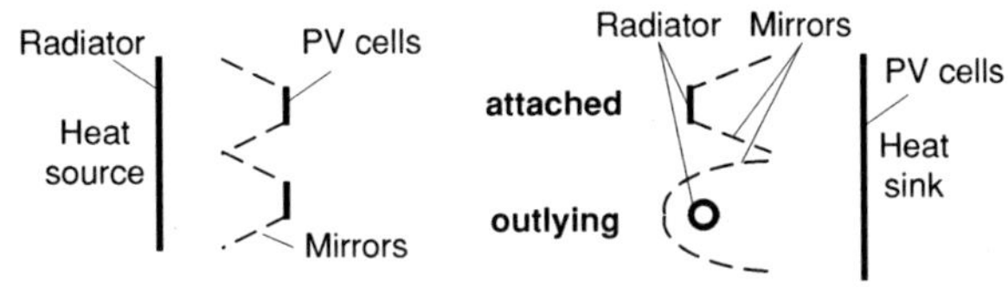

Fig. 6.7 Cavity configurations with concentrators and collimators, concentrator at the PV cell (*left*) and radiator with collimator (*right*)

performance of the systems by an improved spatial and angular radiation distribution. Figure 6.7 shows two types of cavity configurations to direct the radiation. The first configuration utilises a *concentrator at the PV cell*. This could potentially increase the power density and replace expensive PV cell area by cheaper mirror area to reduce costs. The mirrors may extend to the radiator [62], consist of one large concentrator [74] or several small concentrators [75, 76]. An experiment by Fraas et al. pointed out that mirror area decreases the efficiency due to parasitic air conduction and convection and it was concluded that mirror areas should be minimised [77]. On the other hand, this problem could be minimised, if the cavity is evacuated. In the authors view, the role of radiation concentration in TPV cavities has not been fully answered yet.

In the second configuration the *radiator* utilises a *collimator* (Fig. 6.7 on the right). Usually the radiator emits diffuse radiation, whereas most filters and PV cells perform better for a radiation distribution with small zenith angles . Hence, collimator configurations have been proposed to direct this diffuse radiation. Reported research work included one large collimator [78], as well as light-emitting groove radiators consisting of a high emissivity back and cone shaped reflective sidewalls (Fig. 6.7 on the right, attached radiator) [79, 80]. The utilisation of a tubular radiator surrounded by a parabolic mirror has also been proposed for the conversion of radiation into a parallel beam (Fig. 6.7 on the right, outlying radiator) [81].

6.3.4 Radiation Guidance by Total Internal Reflection in Dielectrics

Experiments could demonstrate the *guidance of radiation* by total internal reflection in partially transparent rods from a high-temperature source to the cold PV cell. The rods need not be cooled but should have a temperature gradient across their length. Goldstein et al. used a 200 mm long yttrium aluminium garnet (YAG) rod to guide the radiation to the cell [36, 82, 83]. Experiments performed by the author could demonstrate high power densities, and hence effective radiation guidance, using quartz glass rods with a diameter of 25 mm and an effective length from 100 to 200 mm [31]. The author also proposed the combination of radiation guidance by total internal reflection *and the dielectric insulator concept* using no gap between radiator and cell (Fig. 6.8 left, see also Sect. 6.5.2) [84]. Chubb examined this combined approach theoretically by a model using zinc selenide as a dielectric material [85].

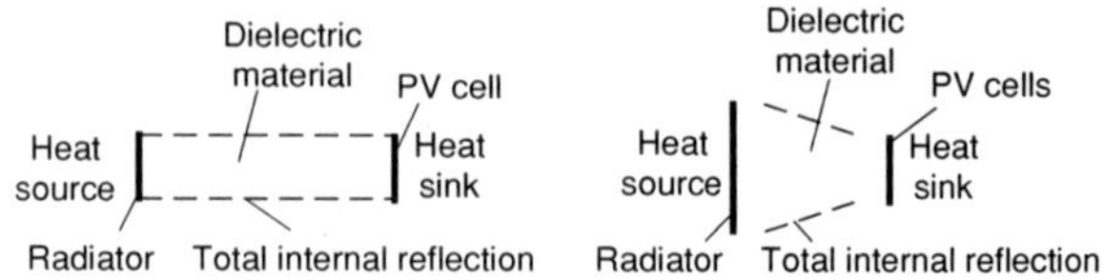

Fig. 6.8 Cavity configurations with dielectric media, dielectric insulator concept (*left*) and dielectric concentrator concept (*right*)

Horne et al. experimentally demonstrated the combination of radiation guidance by total internal *and optical concentration* using quartz glass prisms in the cavity in order to concentrate the radiation 4 times in front of the PV cell [41]. Further research could aim to integrate the three approaches, namely radiation guidance (1), dielectric enhancement (2) and dielectric concentration (3).

6.4 Thermal Insulation Design

In the following text two heat loss paths via the insulation are considered. These paths, visible in the Sankey diagram (Fig. 1.2), are:

- Losses from the radiator to the PV cell via the reflective thermal insulation (see Sect. 6.4.2).
- System losses to surroundings via the thermal insulation

The heat losses via the insulation to the surroundings depend on the surface-to-volume ratio. In particular, smaller systems such as micro generators (see Sect. 6.5.3) have substantial heat losses via this path.

6.4.1 Materials for Thermal Insulation

Figure 6.9 shows a selection of thermal insulation materials collected from various sources [86–90] The half-logarithmic plot shows two groups of materials. The first group has conductivity values higher than still air and the second group utilises the restriction of collisions between air molecules.

For air insulation in physical large chambers at atmospheric pressures, free convection is usually the dominating heat transfer mode. At lower pressures free convection currents disappear and conduction predominates. For heat conduction in *vacuum*, the mean free path is important. The mean free path is defined as the average distance covered by the molecule between successive collisions. For air, at ambient temperature the mean path L_p with the unit cm is given by a simple approximate formula, where p is the pressure in torr[1] (Eq. 6.11) [91]. It can be seen

[1] 1 Torr = 133.3 Pascal; 1 Atmosphere = 760 Torr.

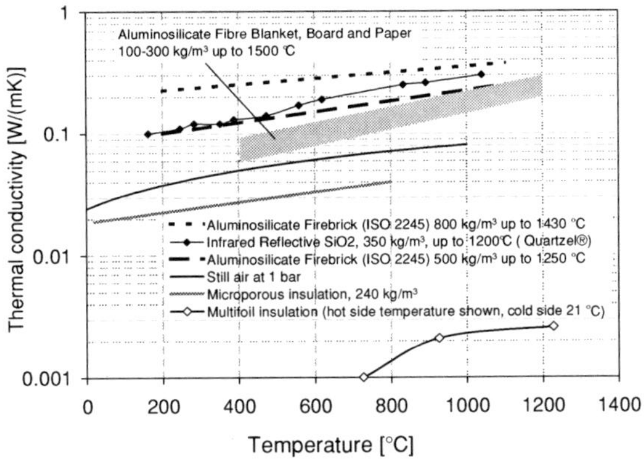

Fig. 6.9 Thermal conductivity versus temperature of various insulation materials

that the mean free path is inversely proportional to pressure and a low pressure results in a large mean free path value.

$$L_p = \frac{5 \cdot 10^{-3}}{p} \tag{6.11}$$

The Knudsen number is a dimensionless number defined as the ratio of the molecular mean free path to a representative physical length scale. Two important extreme cases can be distinguished with respect to the Knudsen number. For Knudsen numbers much smaller than unity (high pressures), the heat conductivity is in the *viscous state*. In this case the totality of molecules is responsible for the heat transfer. The conductivity depends on the temperature, whereas the dependency on the pressure is very weak. On the other hand, for Knudsen numbers much larger than unity (low pressures), the heat conductivity is in the *molecular state*. Here, the individual molecules carry the heat from wall to wall and the conduction heat transfer is proportional to the absolute gas pressure and the temperature difference. Hence, in order to avoid the large conduction heat transfer rate in the viscous state, the gas pressure must be reduced to values where the mean free path is about equal to or greater than the distance between the walls of the containing vessel [92]. For TPV conversion, there is interest in vacuum thermal insulation for applications such as near-field systems (Sect. 6.5.2), radioisotope system in space and solar applications. For the latter, systems with a radiator held with a few contact points in a vacuum chamber have been considered [67, 92–94].

Microporous insulation utilises voids in the material with an average interconnecting pore size comparable to or below the mean free path of air molecules at standard atmospheric pressure. The mean free path of air at atmospheric pressure is easily calculated from Eq. 6.11. The value is 66 nm and pore sizes around or below this value are utilised. Figure 6.9 shows that still *air* at atmospheric pressure has a higher thermal conductivity compared to microporous insulation. Microporous insulation is commercially available (e.g., Microtherm®, Wacker WDS® and Zircar Microsil). For TPV applications, it has been used as an insulation frame between radiator and PV cells [64]. However, currently commercial versions of this insulation are limited to about 1000°C. *Aerogels* utilise also the effect of small pores and they could be another alternative.

Multifoil insulations (MFIs) utilise reflective metal foils separated by spacers, where the space in between is evacuated. MFIs were originally developed for cryogenic applications but are also applicable for high-temperature applications. For radioisotope TPV systems, an MFI using 60 layers of 0.008 mm thick molybdenum foil separated by zirconia (ZrO_2) spacers has been considered [45]. MFIs with molybdenum foil can withstand hot side temperatures up to 1500 K [88].

Other commonly applied high-temperature thermal insulation options that do not utilise mean free path effects, include fibrous and granular insulations, as well as porous firebricks. *Fibre insulation* made mainly from silica and alumina is available in several forms (e.g., board or blanket) for different temperature ranges. One disadvantage in terms of handling is the health and safety issue associated with the fibre dust. Handling of *powder insulation* (e.g., perlite and vermiculite) is often also difficult due to the small particle size. *Insulating firebricks* are less health and safety critical, but they have usually slightly higher conductivities and require machining (Fig. 6.9). For all types, fibre, powder and solid, there is strong dependency between thermal conductivity and porosity (or density).

Another possibility to reduce conductive heat transfer is the replacement of air by an *alternative gas*. For example, the conductivity of krypton, argon and carbon dioxide is lower than air at room temperature and atmospheric pressure [32, 88].

6.4.2 Reflective Thermal Insulation Design

As pointed out in the last section, some cavities do not require mirror areas. For example, the cavity can be entirely covered with PV cells or total internal reflection in dielectrics can be used. However, in most cases a reflective wall is required. The reflective wall must have not only a high reflectivity, but also a low thermal conductivity. A low thermal conductivity insulation of the TPV cavity between the radiator and the PV cell is essential to minimise conduction heat losses via this path. In addition, the reflective thermal insulation must withstand the temperature reached at the radiator (usually higher than 1000°C) and operate next to the PV cell below 100 °C. Hence, the reflective thermal insulation need to be designed to meet these engineering requirements. Aspects, such as the thermal stability and thermal expansion, need to be considered.

Bulk materials with favourable thermal and optical properties are *porous ceramics with a glazed surface* made of alumina and silica. For example, sintered alumina can have high reflectivity values of over 96% in the wavelength range from 0.5 to 2.5 μm. This material is used for laser pumping and lamp design (e.g., Sintox AL) [95]. For this application, the material is designed with a high density and this leads also to high thermal conductivity values. For TPV cavities, low

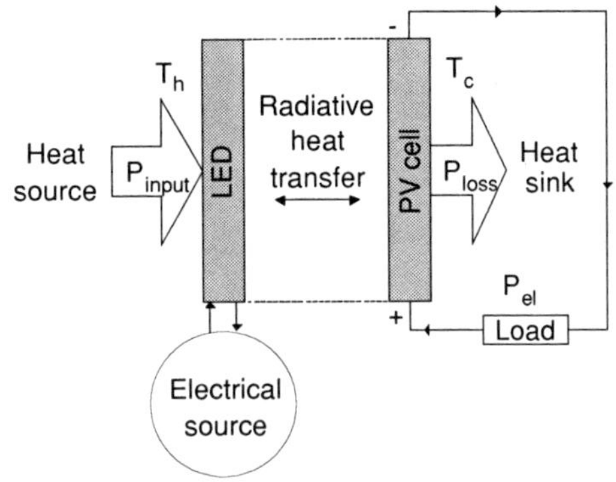

Fig. 6.10 Schematic of thermophotonic (TPX) conversion

density, or low conductivity, reflective porous alumina would be required. Another potentially interesting insulation option could be sintered fused silica fibres, since this material has inherently a high infrared reflectivity and good insulation properties (e.g., Quartzel® up to 1200°C [89]). The long wavelength performance of glazed surfaces is likely to be rather poor due to radiation absorption. Hence, this aspect and vapour pressure limitations would need to be assessed.

Metallic coatings were also studied as a reflector surface. For electric lamp design, reflective coatings consist of aluminium, silver or aluminium–copper (gold colour). These coatings are typically applied by evaporation techniques. For more durable coatings sputtering is used [96]. For example, electrical infrared heaters utilise gold mirrors on quartz glass tubes in order to direct the radiation [97]. A TPV system with a gold coating (1 μm) on glass substrate has been reported [58]. Gold and platinum on specially prepared ceramic foils have been used as reflectors between the radiator and the PV cell [64]. Above 1000°C, titanium in argon atmosphere has been used as a radiation shield [73].

6.5 Novel and TPV-Related Concepts

There are a number of other concepts that are not based on common TPV system designs. The following subsections discuss these novel concepts.

6.5.1 Thermophotonics

The principle of *thermophotonics* (TPX) has been proposed by Green et al. [72, 92, 98–101]. TPX is based on the idea that a light emitting diode (LED) illuminates a PV cell, where the LED operates at a higher temperature than the PV cell (Fig. 6.10). The LED acts as an active radiator that emits photons having a higher energy than the bandgap energy, where the excess energy is provided by a heat

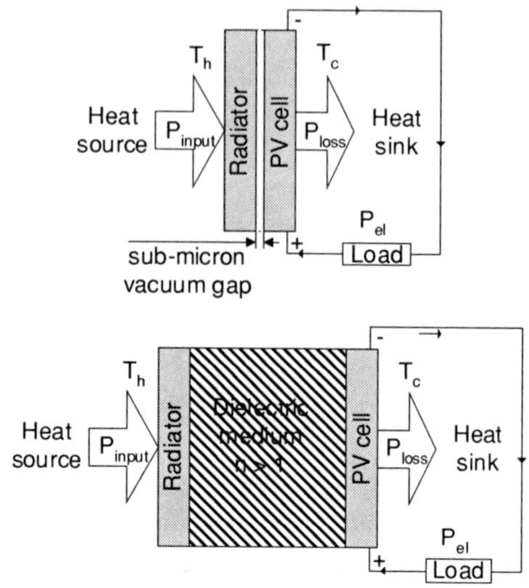

Fig. 6.11 Schematic of near-field (NF) TPV (*top*) and dielectric insulator concept (*bottom*)

source. Ideally, there is only radiative heat transfer between LED and PV cell. A PV cell could then convert this spectrally narrow radiation into electricity to supply the load. If the LED is hotter than the PV cell, then the electric power extracted from the PV cell can be higher than the one injected into the LED. TPX conversion is predicted to have the following advantages compared to TPV conversion: higher efficiency, PV cells could have higher bandgaps and LED temperatures could be lower than TPV radiator temperatures. One major challenge of TPXs is that high performance LEDs for high-temperature operation are not available. Also, the external quantum efficiency of the LED should be near unity. For fabrication of such LEDs, silicon carbide as a semiconductor has been considered.

6.5.2 Dielectric Photon Concentration

The radiative heat transfer rate from the radiator to the PV cell can be considerably increased by two methods. Either the radiator and PV cell are in close proximity (near-field TPV) or the cavity is filled with a dielectric material (dielectric insulator concept). Figure 6.11 shows these two options of dielectric photon concentration. Higher radiative heat transfer rates could result in systems with higher electrical power densities, or alternatively in lower radiator temperatures.

The *near-field TPV concept* utilises a sub-micron vacuum gap and terms such as micron-gap, nano-gap, evanescent waves and photon tunnelling have also been used [102–105]. In the literature, the abbreviation “MTPV defines not only micron gap TPV, but also micro-mechanical TPV systems (see Sect. 6.5.3) [106]. In order to avoid confusion, the terms NF-TPVN and MEMS-TPV are used throughout this book. In the late 1990s research in the area of *near-field TPV*

(NF-TPV) started. The blackbody radiation law can be derived from Maxwell's equations under the assumption of large separation of the hot radiation source and the cold sink [107]. This assumption applies to nearly all TPV systems using typical distances in the centimetre range. However, the blackbody radiation law is not longer valid, when the separation decreases to a sub-micron-gap. This gap allows *partial* coupling of the enhanced radiative heat transfer by n^2 while the vacuum gap thermally insulates. It was concluded that the photon flux is limited by the lowest refractive index n in the photonic cavity (PV cell or radiator) and scales with the minimum refractive index squared [104]. The advantage is that both the PV cell and the radiator can have a high refractive index, so that an enhancement in the order of 10 in the radiative heat transfer rate seems to be feasible [103]. The challenges of this concept include the engineering aspects of a gap in the order of 0.1 μm and the spectral control of radiation [103, 104].

Research on the *dielectric insulator concept* is generally limited [84, 103]. A feasibility experiment using oil as a dielectric material with $n > 1$ at low temperature was reported [103]. The experiment could demonstrate that the radiative heat transfer can be ideally enhanced by n^2, if the cavity is filled with a dielectric medium (for the theory, see Sect. 5.4.6). However, oil is not a suitable material for typical radiator temperatures above 1000°C. A major drawback is also that additional unsuitable conductive heat transfer via the dielectric medium needs to be addressed. Work by the author showed that dielectric solids with increased thickness could be used for the dielectric insulator concept [103]. These solids would also require a low absorption coefficient in the in-band wavelength range, a high maximum operation temperature, a low thermal conductivity and ideally a high absorption in the out-of-band wavelength range. For example quartz glass with a thickness in the order of ten centimetres could be used. Another option could be the use of molten or solid salts. Some salts have a low vapour pressure and they are thermally stable at high temperatures. However, this possibility has not been examined yet.

6.5.3 Micro Generators

Miniature TPV systems with a very small output power have also been developed. They could power electronic gadgets such as notebooks and mobile phones. Micro-machining techniques, based on modified semiconductor device fabrication technologies, are often utilised. Microelectromechanical systems (MEMS) integrate mechanical elements, sensors, actuators and electronics. In the literature the term "power MEMS" was suggested to describe micro-systems that generate power or pumped heat [108].

TPV research focused mainly on micro combustion generators as a battery replacement. Even generators with low conversion efficiencies would be attractive because of the large energy density of hydrogen or hydrocarbon as a fuel compared to batteries. Micro-scale generators experience more difficulties with aspects such as sealing, fabrication, assembly and heat losses (large surface-to-volume ratio) compared to macro-scale systems [68]. Currently, the electrical power output of these generators ranges from the order of milliwatts to watts [56, 68, 102, 106, 108–111].

6.5.4 Blackbody Pumped Laser

Light amplification by stimulated emission of radiation (laser) can be achieved by various methods. The pump energy is usually provided in the form of an electric current or light (flash lamp). However, other less common methods are also feasible. Blackbody pumped laser could be attractive, since they can convert heat directly into laser energy. Solar pumped lasers using a blackbody cavity have been demonstrated. The heat source could not only be solar, but also by other high-temperature sources, such as combustion and nuclear. For an efficient conversion, the laser radiation could be spectrally well matched to the PV cell bandgap. At the moment, heat-to-laser conversion efficiencies have been low, but progress in this field may allow an efficient heat-to-electricity conversion using PV cells. DeShazer reported about experiments using laser rods and PV cells in 2001 [82, 112, 113].

6.5.5 TPV Cascaded with Other Converters

TPV systems operating cascaded with another energy conversion device have also been considered. In such systems the TPV converter may operate in the top or bottom cycle. For example, a gas turbine (top cycle) with the exhaust gas utilised by the TPV converter (bottom cycle) has been examined [60]. Also, wasted heat from the TPV cells (top cycle) could be utilised for a low temperature fuel cell reformer or thermoelectric generator (bottom cycle). If TPV would operate in the top cycle, an increased PV cell temperature would be thermodynamically beneficial (see hot cell operation in Sect. 4.7.2) [72, 114, 115]. Cascaded systems of a TPV converter with another conversion device can be associated with complex system design and are mainly of interest for large power systems, where additional complexity can be justified. It can be concluded that there is generally a limited scope for cascaded operation with other technologies due to the high hot and the low cold side temperature of typical TPV systems.

6.6 Summary

This chapter discussed aspects related to heat transfer and design of the cavity as a core part of an efficient TPV system. The cavity is comprised of a radiator, PV cells and reflective insulation. The reflective insulation, between the radiator and the cell, can be considered to be a critical component. If specific requirements such as high operation temperature, wide spectral mirror range and low conductivity need to be met. Currently, ideal commercial products seem to be not available.

The importance of heat transfer was highlighted in Chap. 5 with special reference to radiative heat transfer in this chapter. The presented ultimate efficiency model showed that control of the radiation spectrum is essential in order to achieve high efficiencies. Even so, not only is spectral control important, but also it is necessary to optimise angular and spatial radiation distribution in the cavity. Finally, more recently proposed concepts were discussed that could improve the performance significantly or open new application fields for TPV conversion.

References

1. Coutts TJ (1999) A review of progress in thermophotovoltaic generation of electricity. Renew Sustain Energy Rev 3(2–3):77–184
2. Coutts TJ, Ward JS (1999) Thermophotovoltaic and photovoltaic conversion at high-flux densities. IEEE Trans Electron Devices 46(10):2145–2153
3. Cody GD (1999) Theoretical maximum efficiencies for thermophotovoltaic devices. In: Proceedings of the 4th NREL conference on thermophotovoltaic generation of electricity, Denver, Colorado, 11–14 Oct 1998. American Institute of Physics, pp 58–67
4. Woolf LD (1986) Optimum efficiency of single and multiple bandgap cells in thermophotovoltaic energy conversion. Sol Cells 19(1):19–38
5. Woolf LD (1985) Optimum efficiency of single and multiple band gap cells in TPV energy conversion. In: Proceedings of the 18th IEEE photovoltaic specialists conference, IEEE, pp 1731–1732
6. Wanlass MW, Emery KA, Gessert TA, Horner GS, Osterwald CR, Coutts TJ (1989) Practical considerations in tandem cell modelling. Sol Cells 27:191–204
7. Horner GS, Coutts TJ, Wanlass MW (1995) Proposal for a second-generation, lattice matched, multiple junction Ga_2/AsSb TPV converter. In: Proceedings of the 1st NREL conference on thermophotovoltaic generation of electricity, Copper Mountain, Colorado, 24–28 July 1994. American Institute of Physics, pp 390–403
8. Caruso A, Piro G (1986) Theoretical efficiency of realistic solar cells intended for thermophotovoltaic applications. Sol Cells 19(2):123–130
9. Iles PA, Chu CL, Linder E (1996) The influence of bandgap on TPV converter efficiency. In: Proceedings of the 2nd NREL conference on thermophotovoltaic generation of electricity, Colorado Springs, Colorado, 16–20 July 1995. American Institute of Physics, pp 446–457
10. Höfler H (1984) Thermophotovoltaische konversion der sonnenenergie (in German), Doctoral thesis. Universität Karlsruhe (TH)
11. Würfel P, Ruppel W (1980) Upper limit of thermophotovoltaic solar-energy conversion. IEEE Trans Electron Devices 27(4):745–750
12. Harder NP, Würfel P (2003) Theoretical limits of thermophotovoltaic solar energy conversion. Semicond Sci Technol 18:151–157
13. Bell RL (1979) Concentration ratio and efficiency in thermophotovoltaics. Sol Energy 23(3):203–210
14. Edenburn MW (1980) Analytical evaluation of a solar thermophotovoltaic (TPV) converter. Sol Energy 24(4):367–371
15. Duomarco JL, Kaplow R (1984) Theoretical estimations of the efficiency of thermophotovoltaic systems using high-intensity silicon solar cells. Sol Energy 32(1):33–40

16. Badescu V (2001) Thermodynamic theory of thermophotovoltaic solar energy conversion. Appl Phys 90(12):6476–6486
17. De Vos A (1992) Endoreversible Thermodynamics of Solar Energy Conversion. Oxford University Press, Oxford
18. De Vos A (1993) The endoreversible theory of solar energy conversion: a tutorial. Sol Energy Mater Sol Cells 31:75–93
19. Baruch P, de Vos A, Landsberg PT, Parrott JE (1995) On some thermodynamic aspects of photovoltaic solar energy conversion. Sol Energy Mater Sol Cells 36:201–222
20. Gray JL, El-Husseini A (1996) A simple parametric study of TPV system efficiency and output power density including a comparison of several TPV materials. In: Proceedings of the 2nd NREL conference on thermophotovoltaic generation of electricity, Colorado Springs, 16–20 July 1995. American Institute of Physics, pp 3–15
21. Zenker M, Heinzel A, Stollwerck G, Ferber J, Luther J (2001) Efficiency and power density potential of combustion-driven thermophotovoltaic systems using GaSb photovoltaic cells. IEEE Trans Electron Devices 48(2):367–376
22. Heinzel A, Luther J, Stollwerck G, Zenker M (1999) Efficiency and power density potential of thermophotovoltaic systems using low bandgap photovoltaic cells. In: Proceedings of the 4th NREL conference on thermophotovoltaic generation of electricity, Denver, Colorado, 11–14 Oct 1998. American Institute of Physics, pp 103–112
23. Bhat IB, Borrego JM, Gutmann RJ, Ostrogorsky AG. (1996) TPV energy conversion: A review of material and cell related issues. In: roceedings of the 31st intersociety energy conversion engineering confernce, IEEE, pp 968–973
24. Yeargan JR, Cook RG, Sexton FW (1976) Thermophotovoltaic systems for electrical energy conversion. In: Proceedings of the 12th IEEE photovoltaic specialists conference, IEEE, pp 807–813
25. Fahrenbruch AL, Bube RH (1983) Chapter 12: Concentrators, concentrator systems, and photoelectrochemical cells. In: Fahrenbruch AL, Bube RH (eds) Fundamentals of solar cells, pp 505–540
26. Chubb DL (1990) Reappraisal of solid selective emitters. In: Proceedings of the 21th IEEE photovoltaic specialists conference, IEEE, pp 1326–1333
27. Shockley W, Queisser HJ (1961) Detailed balance limit of efficiency of p-n junction solar cells. Appl Phys 32:510–519
28. Decher R (1997) Direct Energy Conversion Fundamentals of electric power production. Oxford University Press, Oxford
29. Kittl E (1974) Unique Correlation between blackbody radiation and optimum energy gap for a photovoltaic conservation device. In: Proceedings of the 10th IEEE photovoltaic specialists conference, IEEE, pp 103–106
30. Nelson R (2003) TPV Systems and state-of-the-art development. In: Proceedings of the 5th conference on thermophotovoltaic generation of electricity, Rome, 16–19 Sept. 2002. American Institute of Physics, pp 3–17
31. Bauer T (2005) Thermophotovoltaic applications in the UK: Critical aspects of system design. PhD thesis, Northumbria University, UK
32. Fraas L, Avery J, Malfa E, Wuenning JG, Kovacik G, Astle C (2003) Thermophotovoltaics for combined heat and power using low NOx gas fired radiant tube burners. In: Proceedings of the 5th conference on thermophotovoltaic generation of electricity, Rome, 16–19 Sept 2002. American Institute of Physics, pp 61–70
33. Pierce DE, Guazzoni G (1999) High temperature optical properties of thermophotovoltaic emitter components. In: Proceedings of the 4th NREL conference on thermophotovoltaic generation of electricity, Denver, Colorado, 11–14 Oct 1998. American Institute of Physics, pp 177–190
34. Ferguson L, Fraas L (1997) Matched infrared emitters for use with GaSb TPV cells. In: Proceedings of the 3rd NREL conference on thermophotovoltaic generation of electricity, Denver, Colorado, 18–21 May 1997. American Institute of Physics, pp 169–179

35. Horne WE, Morgan MD, Sundaram VS, Butcher T (2003) 500 watt diesel fueled TPV portable power supply. In: Proceedings of the 5th conference on thermophotovoltaic generation of electricity, Rome, Italy, 16–19 Sept 2002. American Institute of Physics, pp 91–100
36. Goldstein MK, DeShazer LG, Kushch AS, Skinner SM (1997) Superemissive light pipe for TPV applications. In: Proceedings of the 3rd NREL conference on thermophotovoltaic generation of electricity, Denver, Colorado, 18–21 May 1997. American Institute of Physics, pp 315–326
37. Bitnar B, Durisch W, Mayor J-C, Sigg H, Tschudi HR, Palfinger G, Gobrecht J (2003) Record electricity-to-gas power efficiency of a silicon solar cell based TPV system. In: Proceedings of the 5th conference on thermophotovoltaic generation of electricity, Rome, Italy, 16–19 Sept 2002. American Institute of Physics, pp 18–28
38. Bitnar B, Durisch W, Mayor J-C, Sigg H, Tschudi HR, Palfinger G, Gobrecht J (2001) Development of a small TPV prototype system with an efficiency more than 2%. In: Proceedings of the 17th european photovoltaic solar energy conference, Munich, 22–26 Oct 2001. WIP
39. Becker FE, Doyle EF, Shukla K (1999) Operating experience of a portable thermophotovoltaic power supply. In: Proceedings of the 4th NREL conference on thermophotovoltaic generation of electricity, Denver, Colorado, 11–14 Oct 1998. American Institute of Physics, pp 394–402
40. Zenker M (2001) Thermophotovoltaische Konversion von Verbrennungswaerme (in German). Doctoral thesis, Albert-Ludwigs-Universität Freiburg im Breisgau
41. Horne E (2002) Hybrid thermophotovoltaic Power Systems EDTEK, Inc., US, Consultant Report, P500-02-048F
42. Fourspring PM, DePoy DM, Beausang JF, Gratrix EJ, Kristensen RT, Rahmlow TD, Talamo PJ, Lazo-Wasem JE, Wernsman B (2004) Thermophotovoltaic spectral control. In: Proceedings of the 6th international conference on thermophotovoltaic generation of electricity, Freiburg, Germany, 14–16 June 2004. American Institute of Physics, pp 171–179
43. Rahmlow TD, Lazo-Wasem JE, Gratrix EJ, Fourspring PM, DePoy DM (2004) New performance levels for TPV front surface filters. In: Proceedings of the 6th international conference on thermophotovoltaic generation of electricity, Freiburg, Germany, 14–16 June 2004. American Institute of Physics, pp 180–188
44. Fraas LM, Avery JE, Huang HX, Martinelli RU (2003) Thermophotovoltaic system configurations and spectral control. Semicond Sci Technol 18:165–173
45. Schock A Or C, Mukunda M (1996) Effect of expanded integration limits and measured infrared filter improvements on performance of RTPV system. In: Proceedings of the 2nd NREL conference on thermophotovoltaic generation of electricity, Colorado Springs, 16–20 July 1995. American Institute of Physics, pp 55–80
46. Demichelis F, Minetti-Mezzetti E (1980) A solar thermophotovoltaic converter. Sol Cells 1(4):395–403
47. Andreev VM, Khvostikov VP, Khvostikova OA, Rumyantsev VD, Gazarjan PY, Vlasov AS (2004) Solar thermophotovoltaic converters: efficiency potentialities. In: Proceedings of the 6th international conference on thermophotovoltaic generation of electricity, Freiburg, Germany, 14–16 June 2004. American Institute of Physics, pp 96–104
48. Swanson RM (1979) A proposed thermophotovoltaic solar energy conversion system. In Proc of the IEEE 67(3):446–447
49. Doyle EF, Becker FE, Shukla KC, Fraas LM (1999) Design of a thermophotovoltaic battery substitute, 4th NREL confernce on thermophotovoltaic generation of electricity, Denver, Colorado, 11–14 Oct 1998. American Institute of Physics, pp 351–361
50. Wedlock BD (1963) Thermal Photovoltaic Effect. In: Proceedings of the 3rd IEEE photovoltaic specialists conference, IEEE, pp A4.1–A4.13
51. Morrison O, Seal M, West E, Connelly W (1999) Use of a Thermophotovoltaic generator in a hybrid electric vehicle. In: Proceedings of the 4th NREL confernce on thermophotovoltaic

generation of electricity, Denver, Colorado, 11–14 Oct 1998. American Institute of Physics, pp 488–496
52. Guazzoni G, McAlonan M (1997) Multifuel (liquid hydrocarbons) TPV generator. In: Proceedings of the 3rd NREL conference on thermophotovoltaic generation of electricity. Denver, Colorado, 18–21 May 1997. American Institute of Physics, pp 341–354
53. Adair PL, Zheng-Chen, Rose F (1997) TPV power generation prototype using composite selective emitters. In: Proceedings of the 3rd NREL conference on thermophotovoltaic generation of electricity, Denver, Colorado, 18–21 May 1997. American Institute of Physics, pp 277–291
54. Rumyantsev VD, Khvostikov VP, Sorokina O, Vasilev AI, Andreev VM (1999) Portable TPV generator based on metallic emitter and 1.5-amp GaSb cells. In: Proceedings of the 4th NREL conference on thermophotovoltaic generation of electricity, Denver, Colorado, 11–14 Oct 1998. American Institute of Physics, pp 384–393
55. Guazzoni GE, Rose MF (1996) Extended use of photovoltaic solar panels. In: Proceedings of the 2nd NREL confernce on thermophotovoltaic generation of electricity, Colorado Springs, 16–20, July 1995. American Institute of Physics, pp 162–176
56. Yang WM, Chou SK, Shu C, Xue H, Li ZW (2004) Development of a prototype micro-thermophotovoltaic power generator. J Phys D Appl Phys 37:1017–1020
57. DeBellis CL, Scotto MV, Scoles SW, Fraas L (1997) Conceptual design of 500 watt portable thermophotovoltaic power supply using JP-8 fuel. In: Proceedings of the 3rd NREL conference on thermophotovoltaic generation of electricity, Denver, Colorado, 18–21 May 1997. American Institute of Physics, pp 355–367
58. Durisch W, Bitnar B, Roth F, Palfinger G (2003) Small thermophotovoltaic prototype systems. Sol Energy 75:11–15
59. Kushch AS, Skinner SM, Brennan R, Sarmiento PA (1997) Development of a cogenerating thermophotovoltaic powered combination hot water heater/hydronic boiler. In: Proceedings of the 3rd NREL conference on thermophotovoltaic generation of electricity, Denver, Colorado, 18–21 May 1997. American Institute of Physics, pp 373–386
60. Erickson TA, Lindler KW, Harper MJ (1997) Design and construction of a thermophotovoltaic generator using turbine combustion gas. In: Proceedings of the 32nd intersociety energy conversion engineering conference, IEEE, pp 1101–1106
61. Fraas L, Avery J, Malfa E, Venturino M (2002) TPV tube generators for apartment building and industrial furnace applications. In: Proceedings of the 5th conference on thermophotovoltaic generation of electricity. Rome, 16–19 Sept 2002. American Institute of Physics, pp 38–48
62. Sarraf DB, Mayer TS (1996) Design of a TPV Generator with a durable selective emitter and spectrally matched PV cells. In: Proceedings of the 2nd NREL conference on thermophotovoltaic generation of electricity, Colorado Springs, 16–20. July 1995. American Institute of Physics, pp 98–108
63. Schubnell M, Gabler H, Broman L (1997) Overview of European activities in thermophotovoltaics. In: Proceedings of the 3rd NREL confernce on thermophotovoltaic generation of electricity. Denver, Colorado, 18–21 May 1997. American Institute of Physics, pp 3–22
64. Volz W (2001) Entwicklung und Aufbau eines thermophotovoltaischen Energiewandlers (in German), Doctoral thesis, Universität Gesamthochschule Kassel, Institut für Solare Energieversorgungstechnik (ISET)
65. Qiu K, Hayden A (2004) A novel integrated TPV power generation system based on a cascaded radiant burner. In: Proceedings of the 6th international conference on thermophotovoltaic generation of electricity, Freiburg, Germany, 14–16 June 2004. American Institute of Physics, pp 105–113
66. Stone KW, Leingang EF, Kusek SM, Drubka RE, Fay TD (1994) On-Sun test results of McDonnell Douglas' prototype solar thermophotovoltaic power system. In: Proceedings of the 24th IEEE photovoltaic specialists confernce, IEEE, pp 2010–2013

67. (2004) Thermophotovoltaic research, presentation. Space Power Institute, Auburn University, [Online]. Accessed 10 May 2004
68. Nielsen OM, Arana LR, Baertsch CD, Jensen KF, Schmidt MA (2003) A Thermo-photovoltaic micro-Generator for portable power applications. In: Proceedings of the 12th international conference on solid state sensors, Actuators and microsystems, Boston, June 8–12 2003. pp 714–717
69. Chubb D (2007) Fundamentals of Thermophotovoltaic Energy Conversion. Elsevier science, Amsterdam
70. Good BS, Chubb DL (1997) Effects of geometry on the efficiency of TPV energy conversion. In: Proceedings of the 3rd NREL conference on thermophotovoltaic generation of electricity, Denver, Colorado, 18–21 May 1997. American Institute of Physics, pp 487–503
71. Carlson RS, Fraas LM (2007) Adapting TPV for Use in a standard home heating furnace. In: Proceedings of the 7th world conference on thermophotovoltaic generation of electricity, Madrid, 25–27 Sept 2006. American Institute of Physics, pp 273–279
72. Luque A (2007) Solar thermophotovoltaics: Combining solar thermal and photovoltaics. In: Proceedings of the 7th world conference on thermophotovoltaic generation of electricity, Madrid, 25–27 Sept 2006. American Institute of Physics, pp 3–16
73. Aicher T, Kästner P, Gopinath A, Gombert A, Bett AW, Schlegl T, Hebling C, Luther J (2004) Development of a novel TPV power generator. In: Proceedings of the 6th international conference on thermophotovoltaic generation of electricity, Freiburg, Germany, 14–16 June 2004. American Institute of Physics, pp 71–78
74. Schroeder KL, Rose MF, Burkhalter JE (1997) An improved model for TPV performance predictions and optimization. In: Proceedings of the 3rd NREL conference on thermophotovoltaic generation of electricity, Denver, Colorado, 18–21 May 1997. American Institute of Physics, pp 505–519
75. Adachi Y, Yugami H, Shibata K, Nakagawa N (2004) Compact TPV generation system using Al2O3/Er3Al5O12 eutectic ceramics selective emitters. In: Proceedings of the 6th international conference on thermophotovoltaic generation of electricity, Freiburg, Germany, 14–16 June 2004. American Institute of Physics, pp 198–205
76. Fraas LM, Huang HX, Shi-Zhong Ye, She Hui, Avery J, Ballantyne R (1997) Low cost high power GaSb photovoltaic cells. In: Proceedings of the 3rd NREL conference on thermophotovoltaic generation of electricity, Denver, Colorado, 18–21 May 1997. American Institute of Physics, pp 33–40
77. Fraas L, Samaras J, Han-Xiang Huang, Seal M, West E (1999) Development status on a TPV cylinder for combined heat and electric power for the home. In: Proceedings of the 4th NREL conference on thermophotovoltaic generation of electricity, Denver, Colorado, 11–14 Oct 1998. American Institute of Physics, pp 371–383
78. Lindberg E (2002) TPV optics studies–On the use of non-imaging optics for improvement of edge filter performance in thermophotovoltaic applications, Doctoral thesis, Swedish University of Agricultural Science, Uppsala
79. Les J, Borne T, Cross D, Gang Du, Edwards DA, Haus J, King J, Lacey A, Monk P, Please C, Hoa Tran (2000) Interference filters for thermophotovoltaic applications. In: Proceedings of the 15th workshop on mathematical problems in industry, University of Delaware, US, June 1999
80. Modest MM (1993) Radiative heat transfer. McGraw-Hill, New York
81. Regan TM, Martin JG, Riccobono J (1995) TPV conversion of nuclear energy for space applications. In: Proceedings of the 1st NREL conference on thermophotovoltaic generation of electricity. Copper Mountain, Colorado, 24–28 July 1994. American Institute of Physics, pp 322–330
82. DeShazer LG, Kushch AS, Chen KC (2001) Hot dielectrics as light sources for TPV devices and lasers, NASA Tech Briefs magazine, [Online] Available at: http://www.nasatech.com/. Accessed 10 Sept 2004

83. Goldstein MK (1996) Superemissive light pipes and photovoltaic systems including same, Quantum Group Inc., U.S. Pat. 5500054
84. Bauer T, Forbes I, Penlington R, Pearsall N (2005) Heat transfer modelling in thermophotovoltaic cavities using glass media. Sol Energy Mater Sol Cells 88(3):257–268
85. Chubb DL (2007) Light pipe thermophotovoltaics (LTPV). In: Proceedings of the 7th world conference on thermophotovoltaic generation of electricity, Madrid, 25–27 Sept 2006. American Institute of Physics, pp 297–316
86. Guyer EC, Brownell DL (1999) Handbook of applied thermal design. Taylor & Francis, London
87. Baehr HD, Stephan K (2006) Wärme und stoffübertragung (in German), 5th edn. Springer, German
88. Yarbrough DW, Nowobilski J (1999) Section 4.5: thermal insulation. In: Kreith F (ed) CRC Handbook of thermal engineering. CRC Press, Boca Raton
89. (2010) Product: quartzel rigid silica, Saint-Gobain Quartz SAS, France. [Online] Available at: http://www.quartz.saint-gobain.com/. Accessed 28 April 2010
90. Routschka G, Granitzki K-E (2005) Refractory ceramics, in ullmanns encyclopedia of industrial chemistry. Wiley, London, NY
91. Roth A (1990) Vacuum technology, 3rd edn. Elsevier science, Amsterdam
92. Imenes AG, Mills DR (2004) Spectral beam splitting technology for increased conversion efficiency in solar concentrating systems: a review. Sol Energy Mater Sol Cells 84:19–69
93. Khvostikov VP, Rumyantsev VD, Khvostikova OA, Gazaryan PY, Kaluzhniy NA, Andreev VM (2004) TPV cells based on Ge, GaSb and InAs related compounds for solar powered TPV systems. In: Proceedings of the 19th european photovoltaic solar energy conference, Paris, 7–11 June 2004. WIP
94. Andreev VM, Grilikhes VA, Khvostikov VP, Khvostikova OA, Rumyantsev VD, Sadchikov NA, Shvarts MZ (2004) Concentrator PV modules and solar cells for TPV systems. Sol Energy Mater Sol Cells 84(1–4):3–17
95. Mattarolo G (2007) Development and modelling of a thermophotovoltaic system, Doctoral thesis, University of Kassel
96. van den Hoek WJ, Jack AG, Luijks GMJF (2005) Lamps, in ullmanns encyclopedia of industrial chemistry. Wiley, Londen, NY
97. (2006) Infrared emitters for industrial processes, Heraeus [Online] Available at: http://www.noblelight.net/. Accessed 28 Oct 2010
98. Green MA (2003) Chapter 9: thermophotovoltaic and thermophotonic conversion. In: Green MA (ed) Third generation photovoltaics–advanced solar energy conversion. Springer, Berlin, NY, pp 112–123
99. Harder NP, Green MA (2003) Thermophotonics. Semicond Sci Technol 18:270–278
100. Green MA (2001) Third generation photovoltaics: Ultra-high conversion efficiency at low cost. Prog Photovolt: Res Appl 9(2):123–135
101. Tobias I, Luque A (2002) Ideal efficiency and potential of solar thermophotonic converters under optically and thermally concentrated power flux. Trans Electron Devices 49(11):2024–2030
102. Basu S, Chen Y-B, Zhang ZM (2007) Microscale radiation in thermophotovoltaic devices–A review. Int J Energy Res 31(6–7):689–716
103. Baldasaro PF, Fourspring PM (2003) Improved Thermophotovoltaic (TPV) performance using dielectric photon concentrations (DPC), Lockheed Martin Inc., US, Technical Report, LM-02K136
104. DiMatteo R, Greiff P, Seltzer D, Meulenberg D, Brown E, Carlen E, Kaiser K, Finberg S, Nguyen H, Azarkevich J, Baldasaro P, Beausang J, Danielson L, Dashiell M, DePoy D, Ehsani H, Topper W, Rahner K, Siergiej R (2004) Micron-gap ThermoPhotoVoltaics (MTPV). In: Proceedings of the 6th international conference on thermophotovoltaic generation of electricity, Freiburg, Germany 14–16 June 2004. American Institute of Physics, pp 42–51

105. Hanamura K, Mori K (2007) Nano-gap TPV generation of electricity through evanescent wave in near-field above emitter surface. In: Proceedings of the 7th world conference on thermophotovoltaic generation of electricity, Madrid, 25–27 Sept 2006. American Institute of Physics, pp 291–296
106. Xue H, Yang W, Chou SK, Shu C, Li Z (2005) Microthermophotovoltaics power system for portable MEMS devices. Microscale Thermophys Eng 9:85–97
107. Raynolds JE (1999) Enhanced electro-magnetic energy transfer between a hot and cold body at close spacing due to evanescent fields. In: Proceedings of the 4th NREL conference on thermophotovoltaic generation of electricity, Denver, Colorado, 11–14 Oct 1998. American Institute of Physics, pp 49–57
108. Jacobson SA, Epstein AH (2003) An informal survey of power MEMS, the international symposium on micro–mechanical engineering, 1–3 Dec 2003. ISMME2003-K18
109. Yang W, Chou S, Shu C, Xue H, Li Z (2004) Effect of wall thickness of micro-combustor on the performance of micro-thermophotovoltaic power generators. Sens Actuators A 119(2):441–445
110. Yang WM, Chou SK, Shu C, Li ZW, Xue H (2003) Research on micro-thermophotovoltaic power generators. Sol Energy Mater Sol Cells 80:95–104
111. Yang WM, Chou SK, Shu C, Li ZW, Xue H (2002) Development of microthermophotovoltaic system. Appl Phys Lett 81(27):5255–5257
112. Chase L (1991) Solar pumping of lasers. In: Proceedings of the workshop: Potential applications of concentrated solar energy, Solar Energy Research Institute, Cole Boulevard, Golden, Colorado, 7–8 Nov 1990. pp 97–98
113. De Young RJ (1988) Overview and future direction for blackbody solar-pumped lasers, Report, NASA-TM-100621
114. Nagashima T, Corregidor V (2007) An overview of the contributions under cell technologies topic. In: Proceedings of the 7th world conference on the thermophotovoltaic generation of electricity, Madrid, 25–27 Sept 2006. American Institute of Physics, pp 127–128
115. Brougham RP, Whale MD (2001) Feasibility of hybrid the thermophotovoltaic and reformer/fuel cell energy conversion systems. In: Proceedings of the 11th canadian hydrogen conference–building the hydrogen economy, Victoria, Canada, 17–20 June

Chapter 7
Competing Technologies

7.1 Introduction

In general it can be assumed that there exists a potential market for TPV systems anywhere that an electrical power source is required [1]. Hence, in order to identify suitable TPV applications, this chapter reviews other deployed and emerging electricity generating technologies.

In TPV literature, competing technologies discussed have included internal heat engine generators [2–6], solar PV systems [3, 4], electro-chemical cells and direct heat-to-electricity converters. The latter category includes thermoelectric, thermionic and alkali metal thermal-to-electric converters (AMTECs) [2–7]. The three electro-chemical cell types are primary (battery), secondary (rechargeable battery) or tertiary (fuel cell) [2–5, 7]. Other literature also gives details about these technologies [8–13]. These sources indicate that external heat engine generators, and in particular Stirling generators, are another competitor group.

This overview of competing technologies has been limited to a range of technologies. From the various potential technologies to generate electricity only those have been considered which also convert one of the following four sources: solar, nuclear, chemical (or fossil) and waste heat. This has led, for example, to the exclusion of renewable conversion technologies other than solar PV (e.g., wind). Technologies above 1 MW have also not been considered because TPV cell costs are currently high and the largest demonstrated TPV systems have powers in the order of kilowatts. Also conversion technologies above 1 MW have high efficiencies and these efficiencies are currently difficult to achieve with TPV conversion. Only direct heat-to-electricity converters commonly found in recent literature have been considered [2–7]. Some other conversion technologies exist and are discussed elsewhere [9].

The technology options are summarised in Fig. 7.1, which illustrates that the heat sources have to be taken into account when comparing TPV to its competitors. For example, fuel cells compete with TPV systems only for the conversion of chemical energy. On the other hand, external heat engine generators (e.g., Stirling)

T. Bauer, *Thermophotovoltaics*, Green Energy and Technology,
DOI: 10.1007/978-3-642-19965-3_7, © Springer-Verlag Berlin Heidelberg 2011

and direct heat-to-electricity conversion devices (e.g., thermoelectrics) can convert heat regardless of which source.

Power densities (e.g., W/kg, W/m^3, and W/cm^2) and efficiencies are important performance indicators to compare these technologies. The next four sections discuss heat engines, electro-chemical cells, direct heat-to-electricity converters and solar PV systems. Afterwards a comparison of these technologies and TPV technology is made.

7.2 Heat Engine Generators

These generators utilise a two-step conversion process. The first conversion into mechanical energy limits the efficiency. The second conversion from mechanical energy into electricity can be generally efficient with values greater than 95% [5]. The requirement for moving parts can be associated with complex starting, noise, high weight, high complexity, high maintenance and short lifetime compared to TPV conversion. In general, internal and external heat engines can be distinguished [12].

7.2.1 Internal Heat Engine Generators

In an internal heat engine, the heat source is also the working fluid in thermodynamic terms and thus they compete with TPV generators only for combustion systems (Fig. 7.1). Internal combustion engines (ICEs) tend to have few components, although not necessarily moving parts, and therefore higher volumetric and gravimetric power densities compared to external heat engines [12]. ICEs can have an *intermittent* (reciprocating engines) or a *continuous* flow and combustion process (e.g., gas turbine). In intermittent engines combustion occurs during a limited part of the cycle only, so that component temperature exposures tend to be less severe in intermittent compared to continuous engines [12]. This usually results in higher efficiencies but higher pollution (CO, NO_x and unburned H_xC_x) of intermittent engines compared to continuous engines [12, 13]. The flow rate can be higher for continuous engines, so that these engines tend to have higher volumetric power densities [13]. The major *intermittent* heat engines are reciprocating spark ignition (Otto cycle) and compression ignition (Diesel cycle) engines.

Small gasoline generators based on ICEs are commercially available at low prices [5, 14, 15]. Yamaguchi et al. [3, 4] reported system efficiencies of portable generators from 10 to 20% in the range from 1 to 10 kW_{el} and efficiencies below 10% for generators smaller than 1 kW_{el}. Above 10 kW_{el} the efficiencies can exceed 30%, where engines are available from vehicles at low prices and high-power densities [4].

Fig. 7.1 Summary of competing technologies to TPV systems. The figure defines four types of heat sources and five classes of energy converters

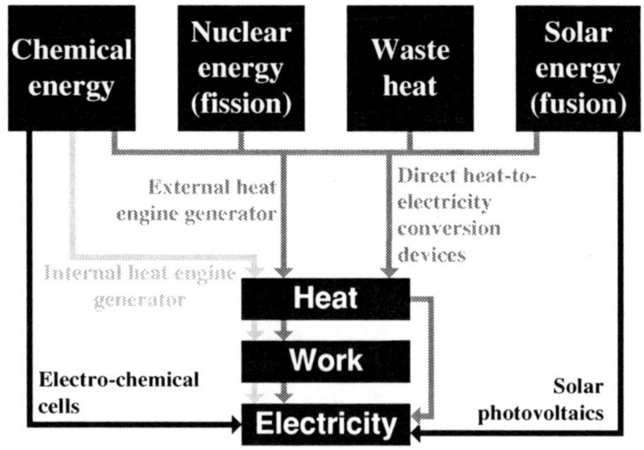

Currently *gas turbine* (Brayton cycle) generators are mainly used in the MW power range, such as for central power stations in conjunction with steam turbines. Gas turbines can operate in a closed or open cycle. The closed cycle configuration results usually in higher efficiencies at the cost of increased complexity [13]. For example, a small commercially available open cycle turbine has a fuel-to-electricity efficiency of about 25% and an electrical output power of 30 kW [16]. Micro turbines as small as 10 W are in a research stage [17].

7.2.2 External Heat Engine Generators

An external heat engine has a separated heat source and working fluid. The use of an external combustor, often associated with heat exchangers for the working fluid, adds usually some complexity and results in lower volumetric and gravimetric power density than ICEs, but allows heat to be derived from all four sources (Fig. 7.1). The major external heat engines apply the Rankine and Stirling cycle.

Steam turbines (Rankine cycle) tend to be used in large centralised power stations. The Rankine cycle can be of closed (e.g., nuclear power plant) or open type (e.g., steam locomotive) [13]. The open cycle requires a constant water source. The closed cycle usually results in complex system designs with moderate efficiencies and low power densities for power outputs considered in this work [13]. Hence these engines are not considered further here.

Stirling engine generators have been researched for a wide power range from artificial heart power to military submarine propulsion. They usually have good part load behaviour, can operate with low noise (free piston engine) and can achieve high efficiencies. Stirling engine generators have been considered for all four heat sources and similar applications as TPV. There are several Stirling combined heat and power (CHP) units emerging commercially. Table 7.1 shows some examples with the electrical efficiency. In general, the CHP efficiency can be high and is not shown. Disadvantages of Stirling engines often include low power

Table 7.1 Examples of performance of stationary CHP units using Stirling engines

Characteristics	WhisperGen (grid-connected) [18]	Stirling 161 [19]	PowerUnit [20]
Manufacturer	WhisperGen	Cleanenergy AB (formerly Solo)	Stirling Biopower (formerly STM Power)
Electrical output power	1 kW	2 –9 kW	43 kW
Efficiency fuel to electricity	~ 10%	25%	27%
Gravimetric power density	7 W/kg	4–20 W/kg	N/A
Volumetric power density	4 kW/m^3	2–11 kW/m^3	N/A

densities [13]. Past challenges in terms of working fluid leakages and seals may or may have not been completely overcomed in current systems.

7.3 Electro-Chemical Cells

There are primary (battery), secondary (rechargeable battery) and tertiary cells (fuel cells), which are discussed in the following two subsections.

7.3.1 Primary and Secondary Cells (Batteries)

Batteries are available from small primary button cells (from 10 J) up to large secondary batteries for underwater propulsion and load levelling (up to 100 MJ) [21]. Advantages include low maintenance (no moving parts), good part load behaviour and efficient charge/discharge performance. Currently, the major primary cell types are zinc-carbon, alkaline-manganese, mercury-oxide, silver-oxide and zinc-oxide, as well as lithium-based cells [22]. The major secondary cell types are lead-acid, nickel-cadmium, nickel-metal hydride and lithium ion. Secondary batteries tend to have higher gravimetric and volumetric *power densities* compared to primary batteries. Secondary nickel-cadmium batteries can have very high-powerdensities with values of around 500 kW/m^3 and above 500 W/kg [23]. This power density can be regarded as high compared to other technologies. Lithium ion batteries have demonstrated high *energy densities* above 0.5 MJ/kg. Nevertheless, the energy density of primary and secondary batteries is low compared to hydrocarbon fuels (see Table 8.6) [23]. Often battery disadvantages include the use of environmentally critical materials, limited shelf life and high electricity costs [23]. For example, common small-scale primary batteries (e.g., AA, 1.5 V, 1,000 mAh) have costs in the order of 100 €/kWh. Also, a shorter recharging time of batteries would be beneficial. An alternative is to replace the battery with a low

Table 7.2 Comparison of the main fuel cell types

	Solid polymer membrane (SPFC)	Alkaline (AFC)	Phosphoric acid (PAFC)	Molten carbonate (MCFC)	Solid oxide (SOFC)
Catalyser	Platinum	Platinum	Platinum	Nickel	Perovskites
Temperature	50–110°C	50–200°C	190–210°C	630–650°C	650–1,000°C
Cell fuels	H_2, CH_3OH	H_2	H_2	H_2, CO	H_2, CO, CH_4
Cell poisons	CO > 10 ppm S > 0.1 ppm	CO_2 is a poison which more or less rules out its use with reformed fuels	CO > 0.5% S > 50 ppm	S > 0.5 ppm	S > 1 ppm
Reformer	External or direct CH_3OH		External	External or internal	External, internal or direct CH_4
Reformer fuels	CH_3OH, alcohols, LPG, gasoline, diesel, jet fuel		Natural gas, alcohols, gasoline, diesel, jet fuel	Gas from coal or biomass, natural gas, gasoline, diesel, jet fuel	Gas from coal or biomass, natural gas, gasoline, diesel, jet fuel
Typical application	Commercial and residential CHP, distributed power, portable power, transport	Transport, space	CHP power generation	Commercial and residential CHP, power generation, ship propulsion, trains	Commercial and residential CHP, power generation, ship propulsion, trains
Advantages	High-power density, rapid start-up and good load-following characteristics, direct methanol cell without reformer	Simple design, cheap electrolyte, high-power density	Advanced technology, commercial system available (e.g., PC25C)	Internal reforming, cheap catalyst, CO as fuel, CO_2 rich fuels	Internal reforming, cheap catalyst CO as fuel, CO_2 rich fuels, impervious to gas cross-over
Disadvantages	CO removal, water management, membrane cross-over and costs	Limited to hydrogen	External reforming, expensive catalyst	Poor start-up, poor load following, high temperature design issues	Poor start-up, poor load following, high temperature design issues

cost hydrocarbon fuel combined with a fuel-to-electricity converter (e.g., TPV, fuel cell). Such fuels can be inexpensive compared to battery costs. For example, current utility gas prices account for euro cents per kWh in Europe. This arrangement has also the advantage of a simpler refuelling compared to the slow recharging process of secondary batteries. Finally, the gravimetric energy density of hydrocarbon fuels is much higher than batteries (see Table 8.6). For these reasons, a hydrocarbon fuel based electricity converter (e.g., TPV, fuel cell) could have advantages in terms of refuelling/charging, costs and energy density compared to batteries.

7.3.2 Tertiary Cells (Fuel Cells)

Fuel cells compete directly with TPV in the conversion of fuel into electricity (Fig. 7.1). Fuel cell research can be found for a wide power range from mW to MW. They can be classified by different criteria including fuel type (e.g., hydrogen or hydrocarbon), fuel processing strategy (external or internal reforming), operation temperature, catalyst material, charge carrier or type of electrolyte. Usually the latter criterion is commonly used for classification. Table 7.2 has been composed from different sources [24–26] and sums up the major fuel cell types and their performance.

One major advantage of fuel cells is the demonstrated high efficiency, which can be around 40–60% based on lower heating values for the major fuel cell types [25, 27]. The efficiency also remains high at part load and over a wide power range. Another advantage is the solid-state operation (e.g., low noise). CHP operation is generally feasible and the operation temperature of the fuel cell determines the heat grade. Hence, solid polymer membrane fuel cells (SPFCs) can only generate low-grade heat. The solid polymer fuel cell (SPFC) is also known as proton exchange membrane (PEM) fuel cell.

SPFCs and advanced space AFCs have demonstrated area power densities (W/cm^2) roughly a factor ten greater than that observed of other fuel cells [27]. This can result in high volumetric and gravimetric power densities. The power density generally increases with increasing power output. For example for automotive applications SPFC stacks have demonstrated power densities of 1,000 kW/m^3 and 700 W/kg [26, 28, 29]. Selecting the operation point of a fuel cell, using its voltage versus current density characteristic, requires a compromise between a highly efficient but large cell and a low efficient but high-power density cell [25].

One key disadvantage of fuel cells is the use of hydrogen, which is commercially not available for the mass market. Also, fuel storage and handling is immature and can be dangerous [5]. The high gravimetric energy density of hydrogen is generally advantageous, whereas the volumetric energy density of 13 MJ/m^3 compares low to oils with values around 40,000 MJ/m^3 (Table 8.6). Compression of hydrogen increases the volumetric energy density. Using a laboratory gas cylinder at 152 bar (15.2 MPa) results in a volumetric energy density of 1,600 MJ/m^3 and higher pressures can increase the density further (e.g., around 5,300 MJ/m^3 at 700 bar). Liquid hydrogen has a volumetric energy density of about 10,000 MJ/m^3 and this is still much lower than hydrocarbon oils. Another storage option is the use of solid-hydride storages (e.g., lanthanum nickel hydride), which are capable of a slightly higher volumetric energy density at the expense of gravimetric energy density, compared to liquid hydrogen. Currently there are challenges with all hydrogen storage technologies [30]. Non-hydrogen fuels require usually internal or external reforming. The difficulties of reforming are discussed elsewhere [25, 26].

SPFCs converting directly methanol without reformer are known as direct methanol fuel cells (DMFCs). There has been some renewed interest since the 1990 s in DMFCs due to advances made in the membrane of the SPFC [26]. DMFCs are attractive, because methanol can be economically and efficiently produced. In addition methanol storage is simple and safe compared to hydrogen. It seems that the first large-scale fuel cell products will be the replacement of secondary batteries by DMFCs (e.g., laptop, mobile phone). In these applications the secondary battery lifetime is already low, so that the DMFCs lifetime is not highly critical. At the time of writing, the DMFCs are entering the market.

7.4 Direct Heat-to-Electricity Conversion Devices

Besides TPV, there are other direct heat-to-electricity conversion technologies including thermoelectric converters, alkali metal thermal-to-electric converters (AMTECs) and thermionic converters. All these technologies use a hot source and a cold sink. The heat flux from the source to the sink can be partly converted into electricity by the direct heat-to-electricity conversion device. Thus important characteristics of these devices are the electricity output per area (power density in W/cm^2) and the efficiency (ratio of electricity output to total heat flux). All technologies share potentially similar solid-state properties, such as simple design, no moving parts, low maintenance, high reliability, good scalability and modularity. Solid-state is a macroscopic statement. In fact, movement or transport of energy must always involve motion of a physical entity with energy [8]. For the direct heat-to-electricity conversion device these are ions (AMTEC), electrons (thermoelectrics, thermionics) or photons (TPV). Also, all technologies can convert heat from any source with a suitable temperature (Fig. 7.1) into direct current electricity. Major parameters where the devices differ from each other include the heat source and sink temperature, the research status, the efficiency and the power density.

7.4.1 Thermoelectric Converter

Electricity generation in a thermoelement is based on the Seebeck, Peltier and Thomson effect. A thermoelectric generator usually consists of a large number of thermoelements connected electrically in series (Fig. 7.2). Of the direct heat-to-electricity conversion devices, thermoelectric generators are the most

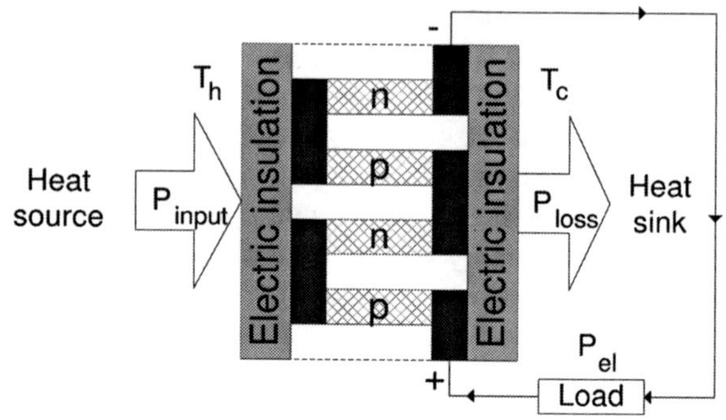

Fig. 7.2 Schematic of the operation principle of a thermoelectric generator. The n-type and p-type semiconductor legs are electrically connected

developed devices and are used in niche market applications such as for space and remote (or non-grid connected) power. Hot side temperatures range from very low (e.g., human body powered wrist watch) up to around 1,300 K (e.g., space applications). Also the power output varies widely from nW to more than 100 kW [31]. Similar thermoelectric modules can be used for cooling as well for power generation from low-grade (e.g., geothermal, ocean power, low-grade waste heat) and high-grade heat (e.g., combustion). In particular the thermoelectric generator using high-grade heat is of interest, since this generator is a closer competitor to the TPV converter.

Calculations of thermoelectric generator efficiency and power density have been discussed by Cobble [32]. The generator can be optimised either for maximum power density or efficiency. Equation (7.1) gives the maximum efficiency of a single-stage generator, which is optimised for maximum efficiency [32]. Equation (7.2) defines figure of merit Z (1/K) for a single-stage converter using one junction material, where S is the Seebeck coefficient (V/K), σ the electrical conductivity (S/m) and k the thermal conductivity (W/mK).

$$\eta_{\mathrm{TE,max}} = \eta_{\mathrm{Carnot}} \cdot \frac{(1 + Z \cdot T_{\mathrm{AV}})^{1/2} - 1}{(1 + Z \cdot T_{\mathrm{AV}})^{1/2} + T_{\mathrm{c}}/T_{\mathrm{h}}} \text{with } T_{\mathrm{AV}} = \frac{T_{\mathrm{h}} - T_{\mathrm{c}}}{2} \tag{7.1}$$

with

$$Z = \frac{S^2 \sigma}{k} \tag{7.2}$$

It follows from Eq. 7.1 that a high efficiency $\eta_{\max}$ requires a large temperature difference from the hot to the cold side T_h–T_c and a high figure of merit Z. Often the dimensionless figure of merit ZT_{AV} is used. There are various n- and p-type materials with a characteristic curve of ZT versus temperature. The material selection is usually based on the ZT performance at the operation temperature. Hence, depending on the application temperature different materials are used. Currently established bulk semiconductor materials have ZT_{AV} values around unity [33–35]. This results in a low heat-to-electricity efficiencies with values of typically less than 5% for established materials. This low efficiency is currently the major drawback of thermoelectric generators.

Long-term experience with thermoelectric generators exists in the area of remote power. For example combustion-based thermoelectric systems can operate in remote oil and gas sites. In this application the low efficiency of 2–3% (fuel-to-electricity) is not critical because of the availability of the fuel. It has been estimated that more than 12,000 combustion-powered thermoelectric generators have been placed into operation for niche market applications [36, 37]. Very remote space missions have utilised radioisotope-powered generators [33, 38]. There has been also a long-term effort in the area of vehicle exhaust gas recovery. At the time of writing there is some renewed interest and several prototypes are being developed. Another application could be the recovery of industrial waste heat. Here, the thermoelectric generators convert a waste heat flux into electricity. It has been pointed out that the low efficiency is not a serious drawback in the conversion of free waste heat and that the capital cost per watt is the decisive economic factor [39]. There have been funding programs for large-scale thermoelectric heat recovery in Japan and the US [33, 40]. Future thermoelectric generators may show higher performance using various concepts to increase the figure of merit Z (e.g., using nano-structured materials) [33, 34]. Other promising concepts include symbiotic and multistage generation. For the former one (symbiotic), the thermoelectric generator is part of a counter-flow heat exchanger [33, 41].

7.4.2 Alkali Metal Thermal-to-Electric Converter (AMTEC)

The AMTEC is a thermally regenerative, electro-chemical device, where an alkali metal flows in a closed loop and cycles its aggregate state between liquid and vapour [5, 8, 42]. Of the alkali metals, sodium has mainly been used. There has also been some work on potassium. Cycling of the alkali metal is obtained using devices such as electromagnetic pumps and capillary wicks [42]. The only remaining moving part in AMTECs using capillary wicks is the enclosed alkali metal (solid-state device behaviour). The electricity is generated in an anode, cathode and electrolyte configuration, where alkali metal ions flow through the electrolyte and electrons bypass the electrolyte (Fig. 7.3) [8, 42, 43]. Usually AMTECs use a beta alumina solid electrolyte (BASE), which is commonly fabricated as a dense microcrystalline ceramic compound consisting of the elements sodium, lithium, aluminium and oxygen [8]. The BASE is a good conductor of sodium ions, but a poor conductor of electrons. The BASE limits the hot side temperature to around 1,300 K, since the electrolyte becomes chemically reactive at higher temperatures [8]. The cold-side temperatures typically range from 400 to 800 K [8]. The two main types of AMTEC cycles, liquid-anode and vapour-anode, are discussed elsewhere in detail [42].

The specific advantages of AMTECs are high efficiency, high-power density and the use of potentially low-cost materials [42]. Electrical power densities of 1 W/cm^2 have been achieved [42] and gravimetric power densities up to 500 W/kg are thought to be feasible [5, 42]. One of the major difficulties of current AMTECs

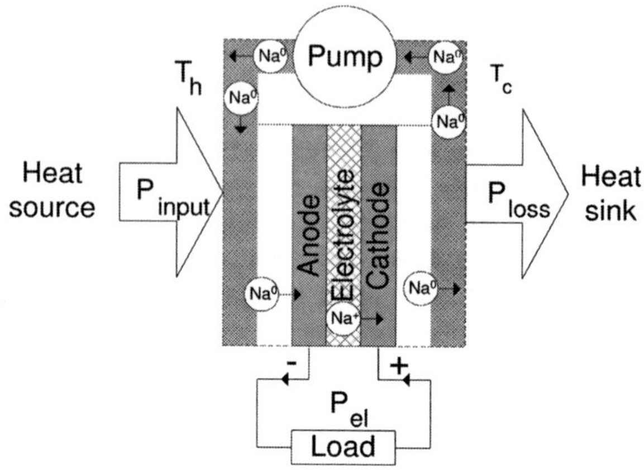

Fig. 7.3 Schematic of the operation principle of an AMTEC. The generator uses the alkali metal sodium, which is ionised in the electrolyte. The figure has been composed from different sources

is the deterioration of efficiency over time. For example an AMTEC, which could only produce about half its initial output power after two years of operation has been reported [42]. Another drawback of AMTECs is that conversion efficiencies larger than 20% of heat into electricity seem to be feasible but this has not been demonstrated [42]. Even though the device is a closed system, handling and safety of sodium can also be critical.

There are several similarities between AMTECs and TPV converters. Both technologies share potentially high-power densities, similar demonstrated and anticipated efficiencies and similar power ranges [5, 42, 44]. For radioisotope powered space applications, both are considered as a possible upgrade for the low-efficient thermoelectric generators at present [44]. A basic literature search indicates that the AMTEC research community is smaller than the TPV community and that AMTEC work has focused on space applications [42]. However, AMTEC work on terrestrial applications, similar to TPV, has also been reported. Applications included remote power, portable power, micro CHP and auxiliary power units (APUs) [45].

7.4.3 Thermionic Converter

In a thermionic converter electrons are emitted (or "boiled") from the heated cathode (also called emitter) and collected by a cooler anode (also called collector). The electrons return to the cathode by means of an external load (Fig. 7.4) [8, 10]. The maximum electron current that an emitting surface can supply per unit area is given by the Richardson-Dushman equation (7.3) [8, 9]. From this equation it becomes clear that the electron flux J increases rapidly with increasing temperatures T_h, and is large for metals with a small work function Φ (e.g., tungsten). The work function of a material is defined as the amount of energy required for an electron with a certain energy level to overcome and escape the binding attractive

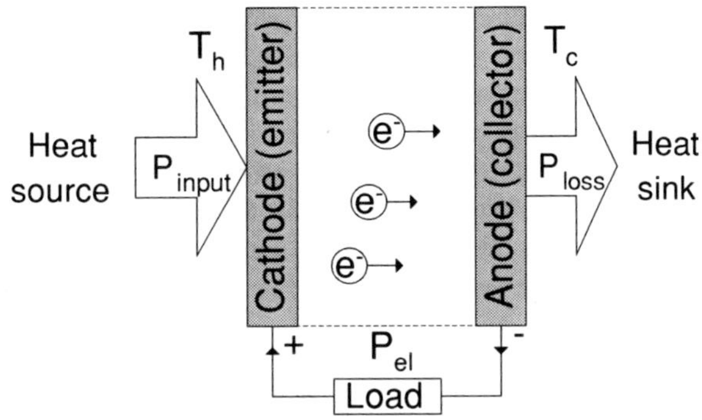

Fig. 7.4 Schematic of the operation principle of a thermionic generator. Electrons are transferred from the hot cathode to the cold anode by the Edison effect

charge (or surface potential) of a material's surface. The constant A is ideally a collection of fundamental constants, where real materials yield lower values [8, 46].

$$J = A \cdot T_{\mathrm{h}}^{2} \cdot e^{-\frac{\Phi}{kT_{\mathrm{h}}}} \text{ with } A = \frac{4\pi e_0 m_{\mathrm{e}} k^2}{h^3} \tag{7.3}$$

One major attraction of thermionic conversion is the high-power density with potential values of 10 s of W/cm^2 [9, 10]. Practically, values in the order of 1–5 W/cm^2 have been achieved [8]. The heat rejection (cold side) temperature of thermionic converters can be high, with temperatures from around 900 to 1,300 K [10, 46]. This should allow cascading with other conversion technologies and the use of small cooling fins for space applications. Another benefit of thermionic generators is the capability to provide peak power above the continuous rating, for a limited time [46].

One challenge in thermionic converter design is that electrons leaving the emitter surface sense the negative space charge in the interelectrode gap and are forced to return to the emitter surface [46]. There have been different attempts to reduce this space charge. One way is the filling of the space with positive ions for neutralisation. Most commonly plasma containing the ionised alkali metal caesium is used. Other approaches are based on closed spaced anodes and cathodes, the use of electric or magnetic fields to conduct the electrons from the cathode to the anode and the use of a grid to accelerate the electron flow [10]. Another major challenge is the high hot side temperature, which is typically around 1,800–2,000 K [46]. At these temperatures major engineering challenges occur which can be associated with high costs. For example cathode evaporation has been reported. This can cause contamination and limits the lifetime [8, 46]. Similar to TPV systems, radiative heat transfer, proportional to T_{h}^{4}–T_{c}^{4}, occurs from the hot (cathode) to the cold (anode) side, whereas this heat transfer is undesirable and degrades the efficiency in thermionic converters.

Thermionic generators have been considered for a wide power range stretching from miniature radioisotope powered converters (mW) to top cycle converters for terrestrial central power stations (MW) [9]. Russia, the US and European countries conducted major thermionic research programs. They focussed mainly on space applications using solar and nuclear heat sources between 1960 and 1990 (e.g., TOPAZ program) [46]. Currently, research in the area of thermionics is limited. The US objectives of thermionic space nuclear reactors have been reported as follows: power range from 10 to 100 kW and beyond, emitter temperature of 2,000 K, collector temperature from 880 to 1,000 K, power density from 3.5 to 13 W/cm^2, efficiency from 10 to 15% and lifetime from 7 to 10 years [47]. More recently there has been also some work on semiconductors using combined thermoelectric and thermionic conversion operating at lower temperatures [48].

7.5 Solar Photovoltaic Systems

Solar concentrator PV is defined here as the direct conversion of (concentrated) solar radiation, whereas solar TPV is the conversion of radiation from an additional intermittent radiator heated by concentrated solar radiation. Even though the operation of solar TPV and solar concentrator PV systems is fundamentally different, both approaches share some similar properties. Both technologies share radiation concentration, the feasibility of CHP operation and similar electrical power densities. Hence, the solar concentrator PV systems can be regarded as a closer competitor compared to non-concentrator solar systems. The combination of solar PV and solar TPV has been proposed [49, 50], but is usually not considered. High solar TPV efficiencies have not yet been demonstrated. Also solar TPV system could have disadvantages in terms of high-temperature engineering and system complexity compared to solar PV systems. Potential advantages of solar TPV include continuous operation using other heat sources (e.g., hybrid system with combustion) and/or thermal storages, low sensitivity to fluctuations in the solar source (e.g., spectrum) and high *potential* efficiency, when compared to solar concentrator PV conversion.

7.6 Summary, Discussion and Comparison with TPV

The discussion in this chapter delivered some insight of the benefits and limitations of various electricity generating technologies in terms of power density (W/m^3, W/m^2, W/kg) and efficiency, as well as other aspects such as heat source (or fuel) flexibility, reliability, lifetime and market status.

Across the conversion technologies, engineering challenges are reported for operation temperatures around or above 1,000°C. These include cracks due to differences in the thermal expansion of materials, contamination degrading the system performance due to material evaporation, thermal insulation losses and radiation effects. Hence, it can be expected that TPV systems will also need to overcome these engineering challenges.

The solid-state technologies discussed in this work can be alternatively classified by their charge carrier: electrons (thermoelectrics, thermionics), ions

Table 7.3 Potential advantages and disadvantages of TPV conversion

Disadvantages	Advantages
Overall system design required	Solid-state
Inter-related research areas	Little or no moving parts
High-temperature engineering	Low noise
High hot side temperature of radiator (T > 1,000°C)	Low maintenance
Low cold side temperature of the PV cell (T < 100°C)	High reliability
High conversion efficiencies not yet demonstrated	Heat source (fuel) variety
Part load behaviour not assessed	Continuous combustion
Currently high PV cell costs	Low pollution, simple ignition
	Long operation time (e.g., industrial waste heat)
	Rapid start-up
	Good scalability (also modular systems)
	Possibility of CHP operation
	Most single components available and tested
	High electrical power density demonstrated

(primary, secondary and tertiary cells, as well as AMTEC) and photons (solar PV, TPV). The technologies considered in this chapter indicate that "electron-based" technologies mainly require materials with suitable properties to increase the efficiency (e.g., figure of merit for thermoelectrics, workfunction for thermionics). On the other hand "ion-based" technologies generally have a requirement for increased lifetime. One may argue that "photon-based" technologies have demonstrated a high lifetime and an intermediate efficiency for solar PV conversion and a similar behaviour may be expected for TPV conversion.

TPV electrical power densities over 2.5 W/cm^2 have been demonstrated [51–53]. The potential TPV power density increases sharply for higher radiator temperature. For example, assuming a PV cell converting 40% of the radiation up to 2.5 μm (e.g., multi-junction InGaAsSb cell) and a 2,000 K blackbody radiator would result in a theoretical electrical power density of about 23 W/cm^2. The power density may be further increased using radiation concentration or high refractive index materials in the cavity (e.g., NF-TPV). Gravimetric power densities for combustion-powered systems greater than 100 W/kg appear to be feasible [11]. Hence, TPV can be regarded as a high-power density technology amongst its competitors.

Table 7.3 sums up the potential advantages and drawbacks of TPV conversion. It can be pointed out that there are possibilities to address most of the disadvantages. For example high-temperature engineering aspects, such as thermo-mechanical and thermo-chemical aspects could be cooperatively solved with experts in the related fields.

There are constraints in the TPV concept due to the limits of the hot and cold side temperature. On the other hand, this limits result in a high-Carnot efficiency. For example a typical system with a radiator temperature of 1,500 K and a cell temperature of 300 K has a Carnot efficiency of 80%. Demonstrated TPV system efficiencies have been moderate. The combustion system efficiencies of up to 8% [54, 55] have been achieved, which can be translated into a heat-to-electricity conversion efficiency of around 10% excluding a combustion efficiency of 80%. In-cavity PV cell efficiencies of around or more than 20% have been reported for InGaAs and InGaAsSb cells [56–59]. For optimised spectral conditions GaSb cell efficiencies of around 30% can be expected (see Sect. 4.6.1). Hence, a 20 to 30% efficient conversion of heat into electricity can be regarded as feasible and competitive, especially if compared to other direct heat-to-electricity converters.

At the current stage, little attention has been paid to the part load operation of a TPV system. Additional secondary batteries for load levelling, modular TPV systems and shutters in the cavity are options to improve the part load behaviour of TPV systems.

Currently, some economic constraints exist for larger TPV systems due to the cell costs. At the moment, small quantities of GaSb cells are commercially available at prices of a few tens €/W, assuming a typical power density of 1 W/cm^2 [60]. Similar to solar PV, the cell price is expected to decrease several orders of magnitude with higher production volumes. It has been pointed out, when expressed in €/W, that the GaSb cells could be almost 100 times less expensive than non-concentrator silicon cells at a given production volume because the production processes of Si and GaSb cells could be similar, but the GaSb cell could generate 1 W/cm^2 instead of 0.01 to 0.02 W/cm^2 [61].

Besides capital cell costs, the operation time is another decisive factor that determines the payback period. Within Europe the annual energy production of PV is 700–2,000 kWh/kW_p per year including Southern European countries with a high insolation such as Spain [62]. For example, a heat recovery TPV system in the high-temperature industry could operate continuously and generate up to 8,760 kWh/KW_p per year. These numbers show that the generated electricity in kWh could be considerably higher (factor 4–13) per installed peak power.

References

1. Ralph EL, FitzGerald MC (1995) Systems/marketing challenges for TPV. Proceedings of the 1st NREL conference on thermophotovoltaic generation of electricity, Copper Mountain, Colorado, US, 24–28 July 1994. American Institute of Physics, pp 315–321
2. Nelson R (2003) TPV Systems and state-of-the-art development. Proceedings of the 5th conference on thermophotovoltaic generation of electricity, Rome, 16–19 Sept 2002. American Institute of Physics, 3–17
3. Yamaguchi H, Yamaguchi M (1999) Thermophotovoltaic potential applications for civilian and industrial use in Japan. Proceedings of the 4th NREL Conference on thermophotovoltaic generation of electricity, denver, Colorado, 11–14 Oct 1998. American Institute of Physics, 17–29
4. Yugami H, Sasa H, Yamaguchi M (2003) Thermophotovoltaic systems for civilian and industrial applications in Japan. Semicond Sci Technol 18:239–246

5. Rose MF (1996) Competing technologies for thermophotovoltaic. Proceedings of the 2nd NREL Conference on thermophotovoltaic generation of electricity, Colorado Springs, 16–20 July 1995. American Institute of Physics, pp 213–220
6. Johnson S (1997) TPV market review. Proceedings of the 3rd NREL Conference on thermophotovoltaic generation of electricity, Denver, Colorado, 18–21 May 1997. American Institute of Physics, pp xxv–xxvii
7. Kruger JS (1997) Review of a workshop on thermophotovoltaics organized for the army research office. Proceedings of the 3rd NREL conference on thermophotovoltaic generation of electricity, Denver, Colorado, 18–21 May 1997. American Institute of Physics, pp 23–30
8. Decher R (1997) Direct energy conversion fundamentals of electric power production. Oxford University Press, Oxford
9. Angrist SW (1976) Direct energy conversion, 3rd edn. Allyn and Bacon, Boston MA
10. Dryden IGC (1975) The efficient use of energy. IPC Science and Technology Press, London
11. Energy sources and systems (1997) in energy-efficient technologies for the dismounted soldier, Chap 3. National Academy Press, Washington, DC [online] Available at:http://www.nap.edu/openbook.php?isbn=0309059348
12. Milton BE (1995) Thermodynamics combustion and engines. Stanley Thornes Publishing Ltd., Cheltenham
13. Theiss TJ, Conklin JC, Thomas JF, Armstrong TR (2000) Comparison of prime movers suitable for USMC expeditionary power sources, Report, Oak Ridge National Laboratory, US, ORNL/TM-2000/116
14. Honda generators (2010) [Online] Available at: http://www.honda-uk.com/. Accessed 5 May 2010
15. Kusko A (1989) Emergency standby power systems. McGraw-Hill, New York
16. Product capstone C30 (2010), capstone turbine corporation, [Online] Available at: http://www.microturbine.com/ (Accessed: 5 May 2010), Chatsworth, US
17. Peirs J, Reynaerts D, Verplaetsen F (2004) A microturbine for electric power generation. Sens Actuators A 113:86–93
18. (2010) Product WhisperGen (grid connected), [Online] Available at: http://www.whispergen.com/ (Accessed: 5 May 2010), WhisperGen Limited, New Zealand
19. (2010) Stirling V161 CHP, [Online] Available at: http://www.cleanergyindustries.com/ (Accessed: 5 May 2010) Cleanergy AB, Sweden
20. (2010) PowerUnit™, [Online] Available at: http://www.stirlingbiopower.com/ (Accessed: 5 May 2010) Stirling Biopower, US
21. Coutts TJ (2001) Thermophotovoltaic generation of electricity. In: Archer MD, Hill R (eds) Clean electricity from photovoltaics, Chap 11, vol 1., Series on photoconversion of solar energy, Imperial College Press, London
22. Deakin RI (2000) Batteries and fuel cells. In: Warne DF (ed) Newnes electrical engineers handbook, Chap. 12. Newnes, Oxford
23. Linden D (1995) Selection and application of batteries. In: Linden D (ed) Handbook of batteries, Chap. 6, 2nd edn. McGraw-Hill, New York, pp 6.1–6.15
24. Srinivasan S, Dave BB, Murugesamoorthi KA, Parthasarathy A, Appleby AJ (1994) Overview of fuel cell technology. In: Blomen LJMJ, Mugerwa MN (eds) Fuel Cell Systems, Chap 2. Plenum Press, New York, pp 37–72
25. Williams MC (2000) Fuel cell handbook, 5th edn. U.S. Department of Energy, Office of Fossil Energy, National Energy Technology Laboratory, DE-AM26-99FT40575
26. Acres GJK (2001) Recent advances in fuel cell technology and its applications. J Power Sources 100:60–66
27. Song C (2002) Fuel processing for low-temperature and high-temperature fuel cells: Challenges, and opportunities for sustainable development in the 21st century. Catal Today 77:17–49
28. The online fuel cell information resource (2010) [Online] Available at: http://www.fuelcells.org/. Accessed 5 May 2010

29. Prater KB (1996) Solid polymer fuel cells for transport and stationary applications. J Power Sources 61:105–109
30. Pinkerton FE, Wicke BG (2004) Bottling the hydrogen genie, the industrial physicist, American Institute of Physics, February/March, pp 20–23
31. Rowe DM, (1994) Chap. 10: Thermoelectric generation. In: Profiting from low-grade heat: Thermodynamic cycles for low-temperature heat sources. Crook AW (ed) Institution of Electrical Engineers
32. Cobble MH (1995) Calculation of generator performance. In: Rowe DM (ed) CRC Handbook of thermoelectrics, Chap 39. CRC Press, Boca Raton
33. Riffat SB, Ma Xiaoli (2003) Thermoelectrics—a review of present and potential applications. Appl Therm Eng 23:913–935
34. Lambrecht A, Böttner H, Nurnus J (2004) Thermoelectric energy conversion - overview of a TPV alternative. Proceedings of the 6th International conference on thermophotovoltaic generation of electricity, Freiburg, Germany, 14–16 June 2004. American Institute of Physics, pp 24–32
35. Rowe DM (2006) General principles and basic considerations. In: Rowe DM (ed) Thermoelectrics handbook: Macro to nano, Chap 1. CRC Press, Boca Raton
36. (2010) Thermoelectric generator. [Online] Available at: http://www.globalte.com/ (Accessed: 8 May 2010) Global Thermoelectric, Canada
37. Hall WC (1995) Terrestrial applications of thermoelectric generators. In: Rowe DM (ed) CRC Handbook of Thermoelectrics, Chap 40. CRC Press, Boca Raton
38. Vining CB (1994) Thermoelectric technology of the future. Presentation, defense science research council workshop, La Jolla, California, 21. July [Online] Available: http://www.poweredbythermolife.com/pdf/Thermoelectric_Technology_of_the_Future.pdf. Accessed 28 April 2010
39. Rowe DM, Min Gao (1998) Evaluation of thermoelectric modules for power generation. J Power Sources 73:193–198
40. Advanced thermoelectric materials for efficient waste heat recovery in process industries (2004) Industrial Technologies Program, U.S. Department of Energy [Online] Available at: http://www.eere.energy.gov/. Accessed 28 April 2010
41. Matsuura K, Rowe DM (1995) Low-temperature heat conversion. In: Rowe DM (ed) CRC Handbook of thermoelectrics, Chap 44. CRC Press, Boca Raton
42. Lodhi MAK, Vijayaraghavan P, Daloglu A (2001) An overview of advanced space/terrestrial power generation device: AMTEC. J Power Sources 103(1):25–33
43. El-Genk MS, Tournier J-MP (2004) AMTEC/TE static converters for high energy utilization, small nuclear power plants. Eng Convers Manag 45:511–535
44. Macauley MK, Davis JF (2001) An economic assessment of space solar power as a source of electricity for space-based activities. Discussion Paper, Resources for the Future, Washington, US [Online] Available at: http://www.rff.org/. Accessed 28 April 2010
45. Oman H (1999) AMTEC cells challenge energy converters. IEEE Aerosp Electron Syst Mag 14:43–46
46. (2001) Overview of the technology, in Thermionics Quo Vadis, An Assessment of the DTRAs Advanced Thermionics Research and Development Program, Chap 3. National Academy Press, pp 15–32, [Online] Available at: http://books.nap.edu/. Accessed 8 May 2010
47. Massie LD (1991) Future trends in space power technology. IEEE Aerosp Electron Syst Mag 6(11):8–13
48. Hagelstein PL, Kucherov Y (2002) Enhanced figure of merit in thermal to electrical energy conversion using diode structures. Appl Phys Lett 81:559–561
49. Davies PA, Luque A (1994) Solar thermophotovoltaics: Brief review and a new look. Sol Eng Mater Sol Cells 33:11–22
50. Woolf LD (1987) Solar photothermophotovoltaic energy conversion, Proceedings of the 19th IEEE Photovoltaic Specialists Conference, IEEE, pp 427–432

51. Carlson RS, Fraas LM (2007) Adapting TPV for use in a standard home heating furnace. Proceedings of the 7th world conference on thermophotovoltaic generation of electricity, Madrid, 25–27 Sept 2006 American Institute of Physics, pp 273–279
52. Fraas L, Groeneveld M, Magendanz G, Custard P (1999) A single TPV cell power density and efficiency measurement technique. Proceedings of the 4th NREL Conference on thermophotovoltaic generation of electricity, Denver, Colorado, 11–14 Oct 1998. American Institute of Physics, pp 312–316
53. Fraas LM, Avery JE, Nakamura T (2002) Electricity from concentrated solar IR in solar lighting applications. Proceedings of the 29th IEEE photovoltaic specialists conference, New Orleans, 19–24 May 2002. IEEE, pp 963–966
54. Volz W (2001) Entwicklung und aufbau eines thermophotovoltaischen energiewandlers (in German), Doctoral thesis, Universität Gesamthochschule Kassel, Institut für Solare Energieversorgungstechnik (ISET)
55. Horne E (2002) Hybrid thermophotovoltaic power systems, EDTEK, Inc., US, Consultant Report, P500-02-048F
56. Wernsman B, Siergiej RR, Link SD, Mahorter RG, Palmisiano MN, Wehrer RJ, Schultz RW, Schmuck GP, Messham RL, Murray S, Murray CS, Newman F, Taylor D, DePoy DM, Rahmlow T (2004) Greater than 20% radiant heat conversion efficiency of a thermophotovoltaic radiator/module system using reflective spectral control. Trans Electron Devices 51(3):512–515
57. Wanlass MW, Ahrenkiel SP, Ahrenkiel RK, Carapella JJ, Wehrer RJ, Wernsman B (2004) Recent advances in low-bandgap, InP-Based GaInAs/InAsP materials and devices for thermophotovoltaic (TPV) energy conversion, Proceedings of the 6th International Conference on thermophotovoltaic generation of electricity, Freiburg, Germany, 14–16 June 2004. American Institute of Physics, pp 427–435
58. Dashiell MW, Beausang JF, Nichols G, Depoy DM, Danielson LR, Ehsani H, Rahner KD, Azarkevich J, Talamo P, Brown E, Burger S, Fourspring P, Topper W, Baldasaro PF, Wang CA, Huang R, Connors M, Turner G, Shellenbarger Z, Taylor G, Jizhong Li, Martinelli R, Donetski D, Anikeev S, Belenky G, Luryi S, Taylor DR, Hazel J (2004) 0.52 eV Quaternary InGaAsSb Thermophotovoltaic diode technology. Proceedings of the 6th international conference on thermophotovoltaic generation of electricity, Freiburg, Germany, 14–16 June 2004. American Institute of Physics, pp 404–414
59. Shellenbarger ZA, Taylor GC, Martinelli RU, Carpinelli JM (2004) High performance InGaAsSb TPV cells. Proceedings of the 6th international conference on thermophotovoltaic generation of electricity, Freiburg, Germany, 14–16 June 2004. American Institute of Physics, pp 345–352
60. Sale Items (2010), JX-Crystals Inc., US [Online] Available at: http://www.jxcrystals.com/. Accessed 28 April 2010
61. Fraas LM, Avery JE, Huang HX (2003) Thermophotovoltaic furnace-generator for the home using low bandgap GaSb cells. Semicond Sci Technol 18:247–253
62. International photovoltaic database (2009), European Comparison [Online] Available at: http://www.sonnenertrag.eu/. Accessed 8 May 2010

Chapter 8
Applications of TPV Generators

8.1 Introduction

8.1.1 Heat Sources

TPV systems can be classified, by the kind of reaction of the heat source, which is either a chemical reaction (rearrangement of the outer electrons of the atoms) or a nuclear reaction (rearrangement of the nucleus of the atoms). Nuclear reactions are of fusion or fission type. This classification results in three major TPV heat sources (Fig. 8.1), which are the combustion of fuels (usually hydrocarbons), solar heat and nuclear sources (radioisotopes and nuclear fission reactors). Terrestrial power generation by nuclear fusion is not considered here further, because effective exploitation seems to be decades in the future.

This book defines a fourth general heat source, waste heat, such as from industrial high-temperature processes. This heat derives mostly from fossil fuel combustion and in some cases from electrical heating, where the major purpose is the provision of process heat. The four sections in this chapter discuss nuclear, solar, combustion and waste heat source application in more detail (Sects. 8.2–8.5).

Hybrid systems consisting of a TPV system combined (not cascaded) with another energy conversion or storage device have also been of interest. Examples include TPV generators combined with a secondary battery providing high peak power [1] or combined with a renewable generator (e.g., solar PV or wind see Sect. 8.4.4) [2]. As already discussed, hybrid systems with more than one heat source (e.g., solar and combustion) are also of interest (see Sect. 8.3.2).

8.1.2 Literature of TPV Applications

In this subsection literature about potential TPV applications is reviewed in chronological order. *Ralph* et al. identified near-term and long-term applications

T. Bauer, *Thermophotovoltaics*, Green Energy and Technology,

DOI: 10.1007/978-3-642-19965-3_8,

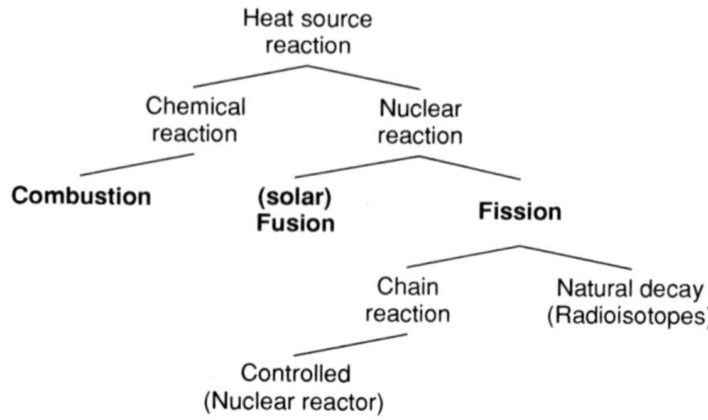

Fig. 8.1 Heat sources of TPV conversion classified by their reactions

for TPV [3]. Near-term markets were characterised by small-scale use, high price and specific TPV advantages. The identified markets were leisure power for applications such as recreational vehicles, boats and cabins (simplicity, quiet and reliable operation and prestige value), battery charges for applications such as military man-packs and portable generators (lightweight), isotope space power for applications such as deep space missions and grid-independent self-powered heaters for applications such gas furnaces and water heater. Potential long-term applications were seen in the vehicles (green car, hybrid electric and military), nuclear (submarines and space reactor) and utility sectors (off-grid, cogeneration and hybrid renewable back-up) [3].

Krist [4] listed potential TPV applications for the gas industry. Self-powered gas heating and cooling devices have been identified as the major applications, such as residential and commercial furnaces, absorption coolers, water heaters, industrial dryers and fireplace power devices (heat circulating and decorative logs). The following advantages were reported: operation during power faults, surplus electricity generation for back-up power, simpler installation (no electrical grid connection required), higher on-site gas consumption (higher gas sales) and energy conservation through on-site generation. In long-term, CHP systems with a minimum gas-to-electric efficiency of 20% and remote power systems for cathodic protection were seen as potential applications. No statement about the power range was made.

Rose [5] gave examples of potential portable TPV applications arranged by their power range: larger than watts to kW (telephones, home electronics, computers, navigational buoy and soldier systems); kW to 10 kW (tools, recreational vehicles, wheelchairs and actuation), 10 to 100 kW (yachts, remotely piloted vehicles, golf carts and electric cars) and larger than 100 kW (advanced radar, spacecraft, electric bus and weapons).

Ostrowski et al. [6] considered the user need, market value, market size, external funding and competing technologies for the assessment. Three classes of applications were identified:

- Near-term: recreational (yacht, RV units) and military.
- Medium-term: commercial (backup power) and remote power (transmitter, cathodic protection and water pumping).

- Long-term: residential (CHP), transportation (low emission fleet and hybrid), electric power (peak loading and grid extension), space mission (satellite).

Johnson [7] predicted that TPV conversion should be advantageous for applications below 5 kW and that there is strong competition above this power. The work defined a hypothetical TPV device in order to assess potential markets. The specifications were assumed as follows: variable power output of 500 to 5,000 W, an efficiency of 10% and a size of $0.6 \times 0.6 \times 0.9$ m^3 (up to 15 kW/m^3). The four application groups identified were recreational vehicles, homes without grid connection, uninterruptible power and military.

Yamaguchi et al. [8, 9] assessed the markets in the area of solar TPV, industrial high-temperature waste heat recovery, micro CHP and portable generators for the commercial Japanese market. Their work selected portable power and micro CHP as the most promising applications. The major requirements for portable generators were identified as high power density, high system efficiency, fuel flexibility, low noise and low price. Competitors (fuel cells, ICE generators, batteries) were also considered and TPV was found to have specific advantages in the power range below 5 kW (especially fuel flexibility, power density and low noise). A micro CHP TPV system with an electrical power of 1 kW and 10 to 20% gas-to-electric efficiency was also modelled in terms of cost and energy savings, where the TPV system supplied the total heat demand using a thermal storage system [8, 9].

Bard presented five major application fields [10]. As an example repeater station for telecommunication was discussed. There are about 6,000 off-grid units with a power of around 50 W in Germany. He discussed costs of a solar PV-battery system and showed that a hybrid system consisting of solar PV, battery and TPV combustion generator could result in more cost-effective solutions. Thermoelectric and DMFC systems were the major competing technologies. The following other application fields were proposed:

- Renewable energy (solar TPV, cogeneration with biomass).
- Small power off-grid supply, <1 kW (environmental monitoring, repeater stations, portable power and backup for PV systems).
- Auxiliary power unit (especially cars, recreational vehicle, sailing boats and trucks, optionally with heat/air conditioning).
- Grid-independent heating appliances (avoidance of electrical grid connection).
- Cogeneration (residential and industrial).

Table 8.1 sums up relevant applications for several competing technologies.

8.1.3 Assumptions of the Application Assessment

At the end of this chapter, an assessment of potential civilian TPV applications is presented (Table 8.5). In the following the necessary assumptions are discussed. As already discussed in Sect. 6.5.5, cascaded systems consisting of a TPV generator and another conversion device are less advantageous. Hence, such systems

Table 8.1 Summary of applications identified from competing technologies

Technology	Applications
Internal heat engine generator (ICE) [11, 12]	Lighting, electronics (e.g., TV, filming and laptop), garden, forest and construction tools (e.g., hedge trimmer, drill, cement mixer, crane and elevator, circular saw, welding), outdoor events (e.g., music, shops), mains backup (home, medical) and auxiliary power for vehicles (e.g., ships, trucks)
Stirling engine generator [13]	Artificial heart power, underwater power unit, space power, remote power sources, military ground power, solar thermal generator and CHP
Battery[14]	Entertainment (lighting, toys and games, photography), vehicle (starting, lighting, ignition, electric/hybrid propulsion, mining, recreational, personal mobility), personal communications devices (portable computers), power tools and backup power (telecommunications, industrial, utility-related)
Fuel cell [15, 16]	1–10 W: camcorder, micro-satellite, palm-top computer, safety lamps and flashlights
	10–100 W: battery re-charger, hand-held power tool, mobile/variable road sign, outdoor/camping supply, portable PC, radio communication and surveillance camera
	100–500 W: domestic gardening equipment, domestic power supply backup, heavy duty battery re-charging, professional power tools and telecommunication field equipment
	1 kW–1 MW: distributed generation (optionally with CHP)
	10–200 kW: road vehicle
	1–10 kW: auxiliary power units (APUs) for vehicles
	Space (satellite) and military (submarine)
Thermoelectric generator [17, 18]	Oil and gas (cathodic protection, supervisory control and data acquisition, offshore), telecom use (relay station, military communication and emergency services), self-powered heating devices, power from vehicle exhaust, power from waste heat
Thermionic generator [19]	Space solar systems (30 to 70 kW), space nuclear reactor (20 kW–MW), terrestrial applications had little attention over the last two decades
Alkali metal thermal-to-electric converter (AMTEC) [20]	Hybrid electric vehicle, portable power (military, battery charger), micro CHP, remote power (lighting, residential), utility power, recreational vehicle, air conditioning power, self-powered furnaces and radioisotope space power
Solar PV [21]	Utility power, recreational vehicles (e.g., boats), remote housing, forest and parks, military, telecommunication, oil and gas (cathodic protection), highway, railroad and marine, agriculture, outdoor lights, refrigerators, computers, lighting, monitoring and instrumentation, remote weather stations, telemetry systems, navigational aids and water pumping

Table 8.2 Summary of efficiency assumptions for the applications assessment. The table shows from left to right the combustion efficiency and the combustion CHP efficiency, as well as the solar, radioisotope and waste heat electrical and CHP efficiency

	Fuel-to-electricity conversion (combustion)		Heat-to-electricity conversion (solar, waste heat, radioisotope)	
	η_{sys} (%)	$\eta_{sys,CHP}$ (%)	η_{TPV} (%)	$\eta_{TPV,CHP}$ (%)
(System cost reduction)	<5	80	<6	100
Demonstrated	5	80	6	100
Near-term	10	80	13	100
Medium-term	15	80	19	100
Long-term	20	80	25	100
(Excluded)	>20	80	>25	100

have not been considered. In the assessment, space and military applications, as well as applications using nuclear sources are briefly discussed to give a comprehensive overview, but have been not discussed in depth, since this book focuses on the commercial and industrial sectors. Only radioisotope generators for space have been included, since this application seems particularly promising. The methodology was to identify first a target power and a target efficiency range. Afterwards the applications are compared by a rating system with four indicators.

The target *power range* of the applications has been expressed on a logarithmic scale in steps of ten. The smallest electrical power currently under consideration for TPV micro generators is in the mW range (see Sect. 6.5.3). A standard upper power range limit of 10 kW has been assumed because competing technologies above this power have high efficiencies (larger than 20%). Also, currently PV cells for TPV systems are only available in limited quantities at high costs. In addition, the maximum demonstrated power of a TPV system has been in the kW range. The absolute upper power range limit has been extended up to 1 MW, if TPV conversion is found to have unique advantages over its competitors. This power range definitions have lead to the exclusion of applications with an electrical power above 1 MW (e.g., centralised power stations).

Efficiency is usually defined as the ratio of the useful output to the total input. The useful output has been assumed either as electricity, or as heat and electricity (CHP mode). Two input modes have been assumed, which were either the product of calorific value and flow rate for a combustion system or a heat flux for all other sources (waste heat, solar and radioisotope). This resulted in four efficiency groups (Table 8.2).

The combustion system efficiencies of around $\eta_{sys} = 8\%$ have been reported [22, 23]. In this work a demonstrated value $\eta_{sys} = 5\%$ has been assumed taking some housekeeping power into account and making cautious assumptions. Applications with lower efficiencies than 5% should allow simple system design at reduced costs. Efficiency targets of 10% (near-term), 15% (medium-term) and 20% (long-term) have been assumed. The medium-term efficiency target has been considered as a standard upper limit for the selected applications. The CHP combustion efficiency has been generally assumed with $\eta_{sys,CHP} = 80\%$ [24].

For waste heat, radioisotope and solar applications, the efficiencies have been derived from the combustion efficiencies by excluding the 20% flue gas loss (combustion efficiencies divided by 80%). Furthermore, it has been assumed that cavity losses and all heat output from the PV cell can contribute to the useful heat output. This can be seen on the CHP efficiency with a value of 100% in Table 8.2 (see also Fig. 1.2). Applications with efficiency requirements higher than the long-term efficiency target ($\eta_{sys} > 20\%$, $\eta_{TPV} > 25\%$) have been excluded from the assessment (e.g., series hybrid electric vehicles, CHP plants above 100 kW power and centralised power stations).

Four *indicator* groups have been identified in this iterative assessment. For each indicator group specific questions have been raised and a rating from 0 to 3 has been introduced. Applications with a 0 rating have been excluded in the iterative assessment. The considered ratings were 1 (negative), 2 (balanced) and 3 (positive).

The first indicator group questions the TPV technology constraints prohibiting the use and the *research and development effort* for a specific application. Three ratings have been defined: negative (1), balanced (2) and positive (3). Factors contributing to a negative rating have been: no TPV system development, complex overall design, operation under part load, operation in a hostile environment (e.g., temperature, humidity or vibration) and high efficiency requirements (15–20%). Positive factors have been: TPV system development of at least one institution, operation partly demonstrated, efficiencies smaller than 5% sufficient and simple overall design.

The second indicator group assesses the benefit of TPV compared to *competing technologies* in a deployed (current technology) or emerging state (likely future technology). The following rating has been used:

0. TPV has disadvantages over one or more other deployed technology.
1. The disadvantages and advantages of TPV and competing deployed technologies are balanced.
2. TPV has advantages over *either* competing deployed *or* emerging technologies.
3. TPV has advantages over competing deployed *and* emerging technologies.

The following factors have been regarded as important for this indicator:

- Noise.
- Reliability, maintenance, dormancy and lifetime.
- Modularity and scalability.
- Efficiency, power density (W/m^3, W/kg).
- Heat source consideration (e.g., fuel storage or flexibility).
- Direct or alternating current power requirements.

The third indicator group *market and cost* has been rated negative (1), balanced (2) or positive (3). The following aspects have been taken into account and each individual positive, balanced or negative rating has been summed up to the overall rating:

- Three is a large potential market and a niche market. The niche market allows for higher costs to launch TPV.
- There is interest from the TPV community (market push).
- There is a market requirement (market pull).
- Long operation hours allow cost-effective operation.
- TPV system costs could match the application.
- Public funding has been available or is seen feasible.

The *human impact* is the fourth indicator. Special attention has been paid to potential primary energy savings (or global CO_2 reductions), but also local human impact factors have been considered. Local factors include low pollution (SO_x and NO_x), low noise, security of supply improvements and user friendliness (e.g., low maintenance). The following rating has been used:

0. TPV operation makes the current human impact worse.
1. TPV operation makes the current human impact neither worse nor better.
2. TPV operation could improve *either* global *or* local human impact factors.
3. TPV operation could improve both global *and* local human impact factors.

8.2 Nuclear Generator

8.2.1 Nuclear Heat Source

Two nuclear fission sources have been of interest for TPV conversion, namely nuclear reactors [25, 26] and isotopes with half-life periods shorter than the naturally occurring isotopes [24, 27–33]. One of the major attractions of nuclear sources is their high gravimetric energy density (MJ/kg) [34–37]. This results in long refuelling periods and makes them attractive for remote area supply (e.g., naval and space). Drawbacks are usually high costs and safety aspects, such as fuel processing and transport, operation, decommissioning, waste disposal and weapon capability. *Radioisotope systems* have been used as long-term source for space, remote area and pacemaker power supply with a wide heat range from mW to kW. Potentially there are over 1,300 radioisotopes [36]. For generators, most commonly Plutonium 238 and Strontium 90 are used. Plutonium 238 has a long half-life of 87 years. Hence, it is preferred for space missions, although it is more costly. The cheaper Strontium 90 with a half-life of 28 years has been used by the former Soviet Union to power remote generators along the coastline [38]. The first publications for TPV nuclear space research date back to the 1980s. Radioisotope temperatures from 1,000 to 1,200°C have been reported in TPV literature [24, 27–33], where the standard Plutonium 238 source defines the upper temperature limit.

The minimum size of a *nuclear reactor* is limited by the critical mass of the fissionable material used. The smallest reactors developed for space exploration had a thermal power in the order of tens of kW heat [35]. At the upper end of the power range are civilian nuclear power stations with GWs of heat. In the reactors, a coolant removes the fission heat. Coolant media include (heavy) water, gases (e.g., helium or carbon dioxide), molten salts and molten metals (e.g., sodium or lead). In a TPV system the coolant would circulate and transport heat from the reactor to the radiator, where molten metal and gas coolants have been proposed for TPV conversion [25, 26]. Gas cooled reactors, such as the currently developmental Pellet Bed Reactors are predicted to achieve radiator temperatures as high as 1,800 K [25].

8.2.2 Nuclear Applications

The major requirements for *space* applications are high reliabilities and the survivability in the space environment, as well as high power densities and high conversion efficiencies. The latter two requirements result in lower launch costs. The only long-duration heat sources are solar and nuclear sources [39], and this is where TPV systems have been considered among other conversion technologies. In unsuitable illumination conditions for PV cells (e.g., near sun and deep space) nuclear sources are used. *Nuclear reactors* for space are considered in a heat range from 10 kW to MWs [19, 35]. Competing conversion technologies include Stirling engine [13], thermionic [19] and thermoelectric generators [35]. TPV space nuclear reactor systems have been proposed but not examined in detail [3, 25].

The major TPV space research focuses on *radioisotope* systems [27–33]. Currently space radioisotope systems utilise thermoelectric converters and for future missions AMTEC, Stirling engine and TPV generators have been considered [20, 40]. TPV radioisotope system efficiencies above 20% are currently projected [24]. TPV systems have the disadvantage of a low heat sink (or PV cell) temperature. This requires large fins to cool the cells in space, or alternatively, operation of the cell at higher temperatures.

The indicator for *technology constraints* and the research and development effort has been rated as negative (Rating 1). Negative aspects were the cell cooling in space by radiation fins, operation in a hostile environment and high efficiency requirements. In terms of *competing technologies*, it has been assumed that radioisotope TPV generator in space would have advantages over competing deployed (thermoelectric generators) and emerging technologies (Stirling, AMTEC). TPV generators could be more efficient than currently used thermoelectric generators. Critical parameters are efficiency and power density (W/kg). Both, TPV and Stirling systems could achieve similar efficiencies, but a TPV system could have a higher power density. A detailed comparison can be found in the literature [41]. Hence, TPV generators should have advantages over both, deployed and emerging technologies (Rating 3). The *market and cost* indicator has been rated positive (Rating 3). The system development aims for a high value niche application with a market push and pull. Other aspects have also resulted in a positive rating, namely the long operation hours, the acceptable system costs and availability of the funding. The replacement of radioisotope thermoelectric generators in space would not result in a major change of the *human impact*. Hence, the human impact has been rated as neither better nor worse (Rating 1).

Terrestrial centralised *nuclear reactor* power stations would require large-sized TPV generators compared to the currently small-sized research generators. Smaller nuclear generators can be found for naval (submarines and aircraft carriers) and remote applications (e.g., repeater station and navigation aids). In the smaller power range Stirling engine generators have been utilised for heat-to-electricity conversion [13]. There are also a few niche-markets for *terrestrial radioisotope* generators, where neither batteries nor combustion systems have sufficient operation times. Examples include very remote power supply (e.g., Polar region) and artificial heart power. These applications differ in their temperature level. For high temperatures, TPV systems are suitable and they would compete with Stirling engine and thermoelectric generators [13].

8.3 Solar Generator

8.3.1 Solar Heat Source

The nuclear fusion in the sun is a powerful and durable heat source. The sun radiates approximately as a blackbody sphere with the radius $r_s = 696 \times 10^6$ m and the temperature $T_s = 5{,}800$ K. The power emitted by the sun is approximately given by $4\pi r_s^2\, \sigma T_s^4$ with a value of approximately 4×10^{26} W, where $4\pi r_s^2$ is the sun surface area and σT_s^4 the Stefan-Boltzmann law. The incident power on the earth is about 2×10^{17} W [42, 43]. This can be compared with the average world primary power demand of around 1×10^{13} W. These values show that conversion of a small share of solar radiation on the earth would be sufficient to meet the world's energy needs. The total solar radiation intensity outside the atmosphere of the earth can be approximately calculated by the equation $r_s^2/r_{se}^2 \sigma T_s^4$, where $r_{se} = 150 \times 10^9$ m is the sun-earth distance and $T_s = 5{,}800$ K is the blackbody temperature. The total value is 1,380 W/m^2 and Fig. 8.2 shows the spectral dependence of the solar radiation. The figure also compares the 5,800 K blackbody spectrum with the standardised AM0 spectrum outside the Earth's atmosphere. It can be seen that the 5,800 K blackbody spectrum closely matches the AM0 spectrum. As solar radiation passes through the atmosphere, it is attenuated by scattering and absorption. The air mass (AM), is the path length through Earth's atmosphere for the solar radiation. Figure 8.2 also shows the terrestrial AM1.5 radiation spectrum [44, 45].

Compared to other heat sources (e.g., combustion and nuclear), the solar source has advantages such as cost-free availability, no pollution and no weight gain through the heat source. The major drawback of the sun as a heat source is the unsteady availability and the low intensity. The incident solar radiation on a surface varies in terms of its spectrum, angle and total intensity depending on factors such as location (e.g., latitude), orientation (e.g., tilt), operation

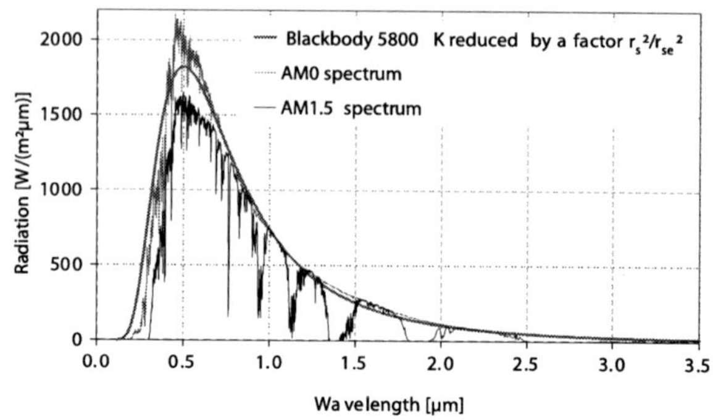

Fig. 8.2 AM0 and AM1.5 solar spectra and 5800 K blackbody radiation at sun-earth distance

environment (e.g., reflected indirect radiation), cloud cover and different cycles (sun, annual, seasonal and daily). At optimum condition (no clouds and optimum tilt angle) the intensity of the terrestrial solar radiation is about 0.1 W/cm^2. The low intensity of solar radiation requires solar concentration in order to achieve suitable radiator temperatures for TPV (>1,000°C).

The principle of *solar concentration* to increase the temperature has long been known (e.g., Archimedes). More recently, solar concentrators have also been extensively used for thermal and electrical systems [46]. In future, solar thermal power plants are predicted to contribute considerably to the electricity supply. A key advantage of these plants is the storage of thermal energy, because heat storage is more economical compared to electrical storage. The same argument can be brought forwards for TPV systems, which could also utilise a high-temperature storage. In general, solar concentrators require locations with a high share of direct sunlight, whereas, locations with a high share of diffuse radiation, are usually not suitable. Concentrating solar systems in high direct insolation areas (e.g., Southern Europe) could supply low insolation areas with diffuse radiation (e.g., Northern Europe) via long distance electric power transmission (e.g., high-voltage direct current lines) [47]. In this way, also low insolation areas could be supplied by solar concentrating systems. Solar concentrators are devices that focus the solar radiation from a large aperture area A_a onto a smaller receiver area A_r facing the sun. Equation 8.1 defines the maximum concentration A_a/A_r, where the sun-earth distance is $r_{se} = 150 \times 10^9$ m, the sun radius is $r_s = 696 \times 10^6\ m$, n is the refractive index and θ_s is the opening half angle of the sun at the Earth's surface. It can be seen that dielectric materials can increase the concentration by a factor n^2. This requires a design with optically coupled absorber and dielectric concentrator [48, 49].

$$\left(\frac{A_a}{A_r}\right)_{max} = \frac{r_{se}^2}{r_s^2} n^2 = \frac{n^2}{\sin^2 \theta_s} = 46448 \cdot n^2 \tag{8.1}$$

The blackbody temperature of the sun surface defines the upper limit of the absorber temperature with a value of about 5,800 K [42]. Experiments demonstrated absorber temperatures as high as 3,000°C [46]. The thermodynamic limit of solar TPV conversion depending on the absorber temperature T_a is given by

Eq. 8.2, where a maximum of 85% at an absorber temperature of 2,478 K occurs, assuming $T_s = 5{,}800$ K and $T_c = 300$ K. It has to be pointed out that this efficiency can be slightly exceeded if the absorber is not a blackbody [42, 50]. It can be also pointed out that the efficiency depends only weakly on T_a at this maximum. Hence, high efficiencies with lower absorber temperatures are feasible.

$$\eta = \left[1 - \left(\frac{T_a}{T_s}\right)^4\right]\left(1 - \frac{T_c}{T_a}\right) \tag{8.2}$$

Solar *concentrator types* can be broadly classified into three categories: non-tracking, single-axis tracking (line focus systems) and two-axis tracking (point-focus systems). For TPV conversion, point-focus systems with a higher concentration achieve suitably high absorber temperatures. TPV work has been reported on dish concentrators [51–55] and a Fresnel point-focus concentrator [56]. Absorber temperatures in STPV systems of 1,350°C have been demonstrated [52, 55]. Concentration levels in the range from 5,000 [57] to 25,000 [58] have been reported. The restrictions in terms of the availability of the sun can be overcome in a solar TPV system by using two strategies, which may also be combined. In the first configuration, the high-temperature heat is *stored thermally* and supplied at times when less or no solar radiation is available. The second configuration is a *hybrid systems* that uses solar radiation *and* an additional heat source (usually combustion) to supply heat for times with no or low availability of solar radiation [53–55, 58–63].

Solar TPV systems have some advantages compared to those TPV systems with other heat sources. Potentially solar systems could operate at very high radiator temperatures close to the thermodynamic optimum at 2,478 K [64–66]. Also the radiator can be completely surrounded by an inert gas or vacuum and this should allow a simple system design (e.g., similar to a light bulb) without challenging high-temperature seals [64–66].

Solar TPV systems may require some additional components. For example, systems using a thermal storage would require a controlled shutter mechanism to minimise heat losses from the storage (see Sect. 2.6).

8.3.2 Solar Applications

The first publications for solar TPV *space* research date back to the 1980s. For space applications, in the majority of cases, solar power is preferred over nuclear power [40]. Applications with intermittent solar radiation (e.g., low earth orbit) usually utilise secondary batteries as electrical storages [40]. These batteries have disadvantages in terms of energy density and lifetime. Attractions of solar TPV for space include the potential of a high efficiency, a high power density and a long lifetime. Solar TPV systems using a high-temperature thermal energy storage have been proposed to replace PV cell/battery space systems [53]. A specific

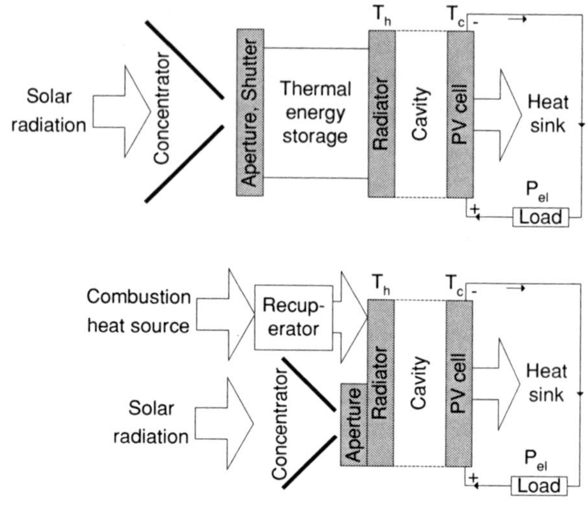

Fig. 8.3 Schematic of a solar TPV system with thermal energy storage (*top*) and a hybrid TPV solar/combustion system (*bottom*)

technological drawback is the requirement for a low PV cell operating temperature resulting in cooling challenges in space. Other technologies considered include Stirling engines and thermionic generators [19].

As discussed earlier, for *terrestrial* applications a suitable climate with a high share of direct solar radiation is usually required. Potential advantages of solar TPV include the high efficiency due to spectral control and the insensitivity to changes in the radiation spectrum, compared to solar concentrator PV systems. At the current research stage, practical solar TPV demonstrators face often difficulties in terms of high-temperature engineering. Solar concentrator systems based on both photovoltaics and Stirling generators have demonstrated high efficiencies [13, 67]. For example, a single multi-bandgap solar PV concentrator cell (GaInP/GaAs/Ge) with an efficiency of about 41% at 240 suns has been reported [68]. Potentially solar TPV systems should also achieve high efficiencies. A solar TPV system was reported that aimed for a solar-to-electric efficiency of 30% [23]. Similarly, in the European research project FULLSPECTRUM solar TPV system efficiencies in a range from 25 to 35% were predicted [69]. At the current stage demonstrated efficiencies are much lower. Also basic PV cell efficiency calculations indicate that a single-bandgap GaSb cell can currently not achieve heat-to-electric efficiencies higher than around 30% in a TPV system (Sect. 4.6.1). These considerations show that currently solar PV concentrator systems using multi-bandgap cells outperform solar TPV systems using a single-bandgap cell. In the long-term solar TPV systems with multi-bandgap cells may be competitive, but the system integration of these cells is currently in an early research stage. Hence, it can be argued that research should not focus on simple solar TPV systems but on other concepts. Solar TPV hybrid or thermal storage systems have other specific advantages compared to solar PV concentrator systems, which can make efficiency considerations secondary (Fig. 8.3).

One possibility is the design of a *hybrid solar-combustion TPV system* (Fig. 8.3 bottom) Such system has been designed using natural gas as a fuel with an electrical output power of around 500 W [23]. This system could also operate in CHP mode. Solar TPV systems using a thermal storage system have been assessed for space applications and may be also applicable for terrestrial use. Thermal storage-based systems could potentially supply heat and power continuously and autonomously with long lifetimes (e.g., no refuelling required and no moving parts). In the short-term such storage and hybrid systems could be used for non-grid connected applications. The niche market would allow for higher costs to launch TPV. Civilian applications may include remote manned (e.g., developing countries) or unmanned power supplies (e.g., relay station, data acquisition, weather stations and navigational aids). For an unsteady load such systems may require an additional electrical storage capability (e.g., secondary battery). In the long-term hybrid solar-combustion TPV systems may be utilised as grid-connected distributed CHP systems [70]. For the thermal storage or hybrid system a long lifetime and a high reliability can be expected. The distributed generation market would be generally very large. Depending on the application and the detailed system design, solar TPV hybrid or thermal storage systems could operate up to 24 h per day.

A prototype hybrid solar/natural gas TPV system has been built. Extrapolation from measurement predicted a solar-to-electricity *efficiency* of 22% and a gas-to-electricity efficiency of about 16% [23]. The project identified a large potential market in the area of grid-connected hybrid solar/natural gas TPV systems with CHP utilisation (e.g., supermarkets, hospitals, hotels, athletic clubs, food processor, restaurants). The ultimate efficiency goals were solar-to-electricity efficiency of 25% and gas-to-electricity efficiency of 20%. For remote (non-grid connected) applications, lower efficiencies could be acceptable. For example thermoelectric converters have been applied in niche market applications with low efficiencies (smaller 5%). Hence, TPV systems with gas-to-electricity efficiencies larger 5% can be regarded as competitive for some niche market applications.

In the small *power range*, hybrid or thermal storage systems could be of interest for unmanned remote applications. In remote, non-grid connected, applications the time interval of site visits and the power requirements are decisive parameters for the selection of a suitable technology (see Fig. 8.4). Flat plate PV cell/battery system or a combustion driven thermoelectric generator may also supply small power applications. Such systems are thought to be less complex compared to a TPV hybrid or thermal storage solar systems. Hence, a minimum power of 100 W can be assumed to justify the more complex hybrid solar TPV approach. Large power installations are likely to allow more maintenance and monitoring, thus permitting the use of deployed diesel engine generators. A maximum competitive power in the order of 10 kW can be assumed.

Compared to other TPV systems, solar TPV storage/combustion systems would need to overcome some additional *technology constraints* compared to simple TPV systems. They include the complex overall design, the operation in hostile environments and the operation under part load for some applications, as well as

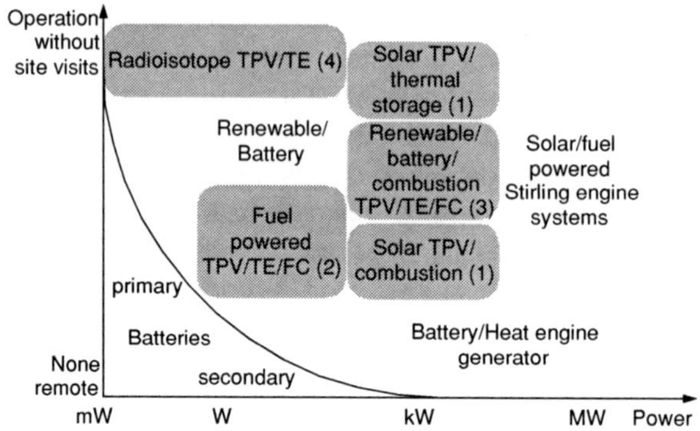

Fig. 8.4 Plot of the identified technology options for remote power supply.The dotted boxes are the areas of interest for TPV generators. The four TPV options considered are solar TPV systems with thermal storage or an additional combustion source (*1*), combustion TPV systems (*2*), combustion TPV systems combined with a battery and a renewable source such as a solar PV or a wind system (*3*) and radioisotope TPV generators (*4*). Thermoelectric generators have been abbreviated with TE, and fuel cells with FC. Applications with a low number of site visits are considered as non-remote

some fundamental work on heat transfer if a thermal storage is included (see Sect. 2.6). Overall, the technology constraints and the research and development effort were rated as negative (Rating 1).

A solar TPV hybrid or thermal storage system is likely to have advantages over both deployed and emerging *competing* technologies (Rating 3). Deployed technologies are solar PV/battery systems, diesel generators and thermoelectric generators. All of them have all their own disadvantages in terms of lifetime and costs (solar PV/secondary battery), maintenance and noise (diesel generator) or efficiency (thermoelectric generator). Another potential system may be the combination of a PV cell, an electrolyser, a hydrogen storage system and a fuel cell. Such system can be regarded as complex and costly. Fuel cell systems operating purely from a fuel tank would have high fuel requirements (large tanks and frequent site visits). Solar/combustion hybrid systems using Stirling engines have also been investigated, but are considered here to suit the power range around or above 10 kW [71].

Market and cost issues can be regarded as positive (Rating 3). There has been at least one funded TPV project [23]. In addition, this application has a potential niche market (e.g., off-grid supply) and a large potential market (distributed generation). Depending on the application and the detailed system design, solar TPV hybrid or thermal storage systems could operate up to 24 h per day. Some colder countries are less suitable, because of the high fraction of diffuse sunlight and lower solar intensity. Nevertheless, for such countries the application offers a large potential export market (e.g., countries with an unreliable or no grid). The *human impact* can be regarded as positive in terms of global and local impact factors (Rating 3).

8.4 Combustion Generator

8.4.1 Combustion Heat Source

Currently, the major part of the world's energy consumption originates from fossil fuel combustion. Combustion as a heat source for TPV conversion has been considered for a wide range of powers. Combustion systems that generate heat as small as about 1 W have been researched [72]. For small power applications, battery substitutes with a thermal power in the order of 100 W have been reported [1, 73–76]. For large power applications, CHP plants based on TPV in the order of 100 kW to 1 MW heat were considered [65, 77].

In general, combustion heat sources achieve suitable radiator *temperatures* for TPV operation (typically 1,000–1700°C). Combustion chamber temperatures depend on the type of fuel (e.g., calorific value, moisture content or aggregate state), type of oxidiser (e.g., air or oxygen), the type of flue gas heat recovery and several other factors (e.g., type of fuel-oxidiser mixing). For TPV, all major methods of flue gas heat recovery have been considered. These include air preheating using recuperator [78, 79] and regenerator devices [80–82], as well as fuel preheating [79] and flue gas recirculation using flame tubes [78, 83]. Lower calorific value fuels tend to generate radiator temperatures at the lower temperature limit of TPV operation. For example for wood powder a radiator temperature of 1,400 K has been demonstrated [65]. The combustion temperature of these fuels may be enhanced by the use of flue gas recirculation [84]. For common fuels combustion temperatures up to 2,500 K are feasible [84]. However at high temperatures, difficulties in terms of thermal engineering (e.g., heat exchanger, thermal insulation) and pollution by thermal nitrogen oxide (NO_x) must be taken into account. In order to avoid excess NOx formation TPV designs usually define a temperature limit. Depending on national regulations maximum temperatures in a range from 1,200 to 1,500°C are reported in the literature [65, 85–87]. Additional, usually more complex, technologies to reduce NO_x emission are available. They include NO_x filters, the replacement of air by oxygen, as well as non-premixed and heat recirculating burners.

Hydrocarbon *fuels* have high gravimetric energy density values (e.g., compared to secondary batteries) [88]. This makes them attractive for portable applications (e.g., petrol in automobiles) and they can be considered as fairly safe (e.g., cigarette lighter in the pocket) [89]. Hydrocarbon combustion has disadvantages in terms of local and global pollution. Excess local pollution including oxides of nitrogen (NO_x) and oxides of sulphur (SO_x) can cause acid rain. Globally the level of carbon dioxide (CO_2) has been found to increase due to the combustion of previously stored (fossil) fuels and the increased atmospheric CO_2 is causing climate change. On the other hand biomass fuels can be regarded as CO_2 neutral.

The most important fuel elements, in terms of heat generation, are carbon and hydrogen (hydrocarbons), which react with oxygen (usually in air) to produce CO_2 and water. For TPV, most commonly *gaseous* fuels, such as methane (or natural

Table 8.3 Approximate volumetric and gravimetric energy density of some selected fuels. Higher calorific values are given. The density of the solids does not include porosities (e.g., powder or granular fills will have lower densities). Primary and secondary batteries are also shown for comparison

Fuel/battery	Type	Density (kg/m^3)	Volumetric energy density (MJ/m^3)	Gravimetric energy density (MJ/kg)
Primary battery	Carbon zinc	2,000	400	0.2
	Alkaline	3,000	1,200	0.4
	Silver oxide	4,500	1,800	0.4
	Lithium	2,500	2,000	0.8
Secondary battery	Lead acid	2,500	250	0.1
	NiCd	3,000	300	0.1
	NiMH	3,000	600	0.2
	Li ion	2,000	1,000	0.5
Gaseous fuel	Hydrogen	0.09	13	142
	Methane	0.7	40	56
	Natural gases	0.7–0.9	33–43	41–54
	Air (for comparison)	1.3	–	–
	Propane	1.9	95	50
Liquid fuel	Oils (e.g., diesel, petrol)	790–970	33,000–42,000	42–47
	Liquefied Propane	510	25,500	50
	Methanol	790	18,200	23
Solid fuel	Coal (lignite—anthracite)	1,100–1,800	28,000–67,000	26–37
	Dry wood	400–900	8,000–18,000	~20

gas), propane or butane, have been used. Combustion of *liquid* fuels usually requires a fuel feed system (e.g., pump) and atomiser, which can make system design more complex compared to gas fuelled systems (gas system can operate solely from a pressurised tank). For military applications, portable TPV systems using logistic liquid hydrocarbon fuels (e.g., kerosene, diesel) have been designed [79, 83, 90, 91]. *Solid* fuels are often associated with sophisticated combustion techniques. Nevertheless, work at the Solar Energy Research Center (SERC) in Sweden focused on cogeneration using wood powder [65]. This fuel diversity demonstrates that, as long as the radiator achieves a suitable temperature, TPV systems can operate with any fuel.

The *chemical storage of energy in hydrocarbon fuels compared with battery system* is of interest for TPV systems using hydrocarbon combustion as a heat source. Hydrocarbon fuels are advantageous in terms of availability, long-term storage and transport characteristics. Liquid and gaseous hydrocarbon fuels share high gravimetric energy densities (Table 8.3) with values more than 100 times higher compared to secondary batteries [88]. Some gaseous fuels can be liquefied in order to achieve similar volumetric energy densities to oils (Table. 8.3) [88, 92–94].

These liquid fuels (diesel, kerosene, liquefied propane) with both high volumetric and gravimetric energy density have been utilised for portable TPV applications [83, 88, 91, 95, 96].

In the majority of cases TPV combustion systems have utilised hydrocarbon fuels, although hydrogen has also been burned in converters with small powers [73]. Hydrocarbon fuels are widely available, can be easily stored, transported and recharged. Additionally they have high gravimetric energy densities [88]. Competing technologies for hydrocarbon-powered TPV generators are mostly batteries in the smaller power range and ICE generators in the larger power range. Fuel cells are emerging for these applications and can be considered as a major competitor.

Williams et al. [76] demonstrated the *direct conversion of flame radiation* with a PV cell without using a radiator. The infrared radiation of hydrocarbon flames is composed of major spectral emission bands at around 2.7 and 4.4 μm [97, 98] and greybody radiation from carbon particles within the flame (soot) [76, 99]. The ratio of radiative heat transfer to the total heat released by the combustion depends on fuel type and burner design as well as other parameters. For natural gas this ratio can reach values as high as 30% for modern industrial burners [99] and Gaydon stated a ratio of 2–20% in an older book [100]. The flue gas can also contaminate optical elements and this can lead to degradation of the system performance. In particular dirty fuels (e.g., liquids) can cause contamination. It can be concluded that the direct conversion of flame radiation is often unsuitable for TPV conversion because of the small ratio of radiant heat to the total combustion heat, the unsuitable radiation spectrum and contamination aspects. Therefore, means to increase the radiation intensity and to tailor the spectrum are required [101]. Nevertheless, the direct conversion of flame radiation with PV cells is very illustrative to show the principle of the TPV concept.

Radiant burners have been developed for both lighting and heating applications. The historical development of spectrally selective radiant burners *for lighting* is of interest, since the mechanisms applied to enhance visible radiation for lighting can also be applied to enhance near-infrared radiation for TPV conversion. In 1826, Thomas Drummond heated a monolithic block of calcium oxide (limelight) to incandescent temperature by flame impingement using a hydrogen/oxygen flame, where the calcia block could enhance visible radiation [101]. Subsequent work focused on different radiator materials and geometries [101]. In the early 1890s, Carl Auer (Baron von Welsbach) perfected both the material compositions and the geometrical structure by using fibres of around 10 μm in diameter consisting of 99.3% of thoria (ThO_2) and 0.7% of ceria (Ce_2O_3) by weight. This structure, known as the Welsbach mantle, continues to be the most efficient converter of gas combustion heat to visible light and is still being used for example in camping lanterns [101, 102]. The Welsbach mantle is optically thick in the visible spectral range and optically thin in the infrared range (1–8 μm) [102]. For most spectral regions the combustion flame is also optically thin, except some spectral bands mainly at around 2.7 and 4.3 μm, so that the Welsbach mantle emits most radiation in the visible range where the mantle is optically thick. The physically small fibre diameters can achieve temperatures close to that of the flame and the combustion products, because their small physical size allows high heat transfer rates. The thin fibres are tolerant to thermal stress and can be heated up quickly [101]. However scaled up Welsbach mantles have proved to be fragile [103].

Radiant burners for heating can be broadly classified into direct and indirect radiant burners. Indirect-fired burners spatially and optically separate the combustion and heating zone, whereas direct-fired burners do not separate these two zones. Alternatively, combustion processes can be classified by the premixing mode of oxidiser and fuel. Premixed burners mix oxidiser and fuel prior to the combustion, whereas non-premixed burners simultaneously mix and burn oxidiser and fuel.

Direct radiant burners may operate on flame impingement or utilise porous structures in the combustion zone. For TPV, physically small porous structures that are of interest are those that can be designed to achieve high temperatures and convert fuel efficiently into radiation.

The physically small porous structures have the advantage that they are in close contact to the flame [104]. High heat transfer rates between combustion gas and these structures are achieved. For this reason, these porous structures reach a high temperature. On the other hand, indirect radiant burners experience necessarily larger temperature gradients from the combustion gas to the radiator. For direct radiant burners, materials considered are typically *oxide type ceramics.* TPV radiant burners have been designed so as to use a modified Welsbach mantle, which consists of ytterbium and erbium oxides. Generally suitably high-temperature operation (more than 1,700 K) and highly selective spectral emission was achieved [103]. However, the modified Welsbach mantle also suffered from scaling and fragility difficulties. Hence alternative structures have been examined including composite fibres and foams. These structures can overcome the scaling and fragility issues of the Welsbach mantle but at the expense of spectral selectivity. Direct radiant burners using *metals* (e.g., wires) have been used in TPV systems. Metals are generally limited to lower temperatures due to oxidation in the flame atmosphere compared to ceramic oxides. One disadvantage of all direct radiant burners is that PV cell surfaces would be contaminated from combustion products. Hence, usually additional means are required to protect the cells. This is usually achieved using transparent shields (e.g., fused silica). No TPV work on the long-term contamination of these shields by the combustion products could be found in the literature, although impurities in these products could affect the shield transparency. Another drawback of direct radiant burners is that at least some unsuitable flame radiation is inherently in the emitted spectrum (at 2.7 and 4.3 μm). Typical PV cells cannot convert this radiation. Furthermore, an optically thin radiator used in direct radiant burners can usually not be used for other heat sources than combustion. At the moment, it seems that there is no commercially available direct radiant burner with a suitable performance for TPV conversion (high efficiency and selective spectrum).

Indirect radiant burners spatially enclose the combustion zone and have been used in TPV systems. These burners utilising gaseous hydrocarbon fuels are commercially available for space and industrial heating applications and are of metal or ceramic type. Ceramic burners are usually of interest for TPV because of their higher operation temperature. These burners typically have a tubular form in various shapes including U, W, P, double P, A or single tube [105]. Single tube burner types include straight-through, single ended recuperative or single ended recirculating recuperative [99, 106]. For TPV the latter one made of silicon carbide (SiC) has typically been utilised. These burners were either commercially acquired [78] or custom-designed [83]. Additional coatings have been applied in order to achieve a spectrally selective emission. The various selective radiator options are discussed in Chap. 2 [104].

Another major engineering aspect is the recovery of flue gas heat to preheat the combustion air. The two conventionally accepted methods of recovering the flue gas heat are recuperation and regeneration. Regeneration, or regenerative firing, is based on cyclical firing of paired combustion systems with thermal energy storage For TPV conversion, regenerators are less suitable because they operate unsteadily and they are usually utilised for larger sized systems. Potentially, a single (steady) burner with a rotary regenerative heat exchanger could be used.

The major recuperator types are cross, parallel and counter flow. For TPV conversion typically counter flow recuperators are used and several designs have been investigated. The counter flow recuperator has different temperature zones. The recuperator in the high-temperature zone is necessarily made of ceramics such as silicon carbide and cordierite. In the low-temperature zone, simpler construction materials, such as stainless steel can be used. Using efficient recuperators about 80% of the chemical energy can be converted to radiant energy (see also energy balance in Fig. 8.2) [24]. Simulated and experimental efficiencies of an optimised counter flow heat exchangers exceeded 90% [86, 107]. Some TPV applications focus on heating rather than a high electrical efficiency. Here, the recuperator requirements are moderate or such system may not require a recuperator at all [108]. Also other arrangements are feasible. For example Qiu et al. proposed a cascaded radiant burner with two temperature zones using silicon cells at the high-temperature zone and GaSb cells at the low-temperature zone [109].

Complete TPV systems require not only the burner and PV cell converter but also several auxiliary parts. They include components for air and fuel conditioning (e.g., pumps, blowers, atomisers), as well as burner ignition and control. Portable systems require a fuel tank and components to remove the heat from the PV cell to the ambient (e.g., pumps, air heat exchanger). For example CHP systems may include a thermal storage tank for hot water. These examples show that there is a larger developmental step from a lab demonstrator to an autonomous operating system.

The following five subsections focus on different combustion applications in the areas of portable power, uninterruptible power supply (UPS), remote power, transportation and combined heat and power (CHP).

8.4.2 Combustion Application: Portable Power

TPV research and development of portable power generators included a multi-fuel (diesel, kerosene) generator with an electrical power of 500 W [79, 83, 91, 110], a propane generator with 100–150 W [111], a portable generator with 5–20 W [112], a battery substitute with 20–25 W [1, 113] and a battery charger combined with a torch with about 1–3 W electrical power [88]. More recently the power range was extended to even lower electrical powers (see Sect. 6.5.3). Such small power systems have not been considered in this assessment because of their early research stage. Hence, the power range of interest has been defined from 1 W to 1 kW including the referred TPV concepts. Batteries are predominant in the lower power range and can be regarded superior below 1 W. Portable heat engine generators have advantages in the higher power range and can be regarded superior above 1 kW. Yamaguchi and co-workers also identified an upper limit of 1 kW [8, 9].

In the US, there has been military interest for portable power generators from the early phases of TPV research up to the time of writing [114–117]. The relevant advantages of TPV for military applications are quiet operation, high power densities, no moving parts, fuel flexibility, capability of CHP operation, tolerance to low temperature, simple start-up, excellent dormancy and direct current output [3, 6, 7, 10, 12, 13, 20, 117, 118]. Challenges were seen in the low demonstrated efficiency, the thermal signature due to the high-temperature operation, sensitivity to temperature changes (temperature control required), poor system experience, radiator ruggedness and up-scaling [117].

Yamaguchi et al. concluded that civilian portable generators are one of the most promising TPV applications [8, 9]. Portable generators with an intermediate power from 10 to 100 W (within the examined range from 1 W to 1 kW) and with a long operation time can be regarded as particularly promising because of low competition from other technologies. Example applications are battery chargers, lighting equipment, portable electronics (e.g., laptop, TV and filming), power tools (e.g., garden, agriculture, forest and construction), camping, temporary outdoor events (e.g., shops and music) and mobile road signs [5, 15, 16].

High conversion efficiencies are desirable in order to achieve long refuelling intervals. Portable hydrocarbon-powered TPV systems with efficiencies (fuel-to-electricity) of a few percent can be superior to current primary and secondary batteries in terms of the overall energy density due to the high energy densities of hydrocarbon fuels (MJ/kg, MJ/m^3) [88]. A minimum efficiency target of 5% has been assumed here, accounting for some weight of the TPV converter in addition to the fuel and assuming that improved energy densities compared to batteries are of interest.

The indicator for *technology constraints* and the research and development effort has been rated as balanced. Several institutions have, at least partly, demonstrated TPV operation for this application [79, 83, 91, 110]. However, a general-purpose generator would need to operate under various operation conditions (part load, hostile environment). Efficient part load operation could be possible using a TPV system in on/off mode and an additional secondary battery, but research on this issue is limited.

Major *competitors* are secondary batteries in the small power range and ICE generators in the large power range. ICE generators have disadvantages in terms of high maintenance, starting in the cold and noise. Major disadvantages of secondary batteries are low lifetime, poor dormancy, slow recharging and limited capacity (or low volumetric and gravimetric energy densities). Fuel cells (emerging) have the advantage of a high efficiency and a good part load performance. On the other hand, TPV generators are predicted to be superior in terms of power density, lifetime and fuel flexibility, compared to fuel cells [8, 9]. It has been concluded that portable TPV generators have some advantages over deployed (secondary battery, ICE generator) and emerging technologies (fuel cell).

There are niche *markets* in portable power that have special requirements for low maintenance, low noise, low weight and long operation time. The entire battery market and heat engine generator market is potentially a large long-term market. In the TPV community there has been interest in military and civilian portable power generators (positive market push). There is also generally a market requirement (positive market pull), since currently secondary batteries often limit the system operation time (e.g., laptop). Funding for civilian portable generators may be regarded as unlikely, but overall market issues have been rated positive. The local *human impact* has been rated positive. Batteries can be regarded as bulky and heavy and difficult to refuel. Compared to heat engine generators, TPV generators should be less noisy and the continuous combustion is generally cleaner. Major CO_2 saving can be not expected for this market and this resulted in a negative global human impact. Hence, overall the human impact has been rated balanced.

8.4.3 Combustion Application: Uninterruptible Power Supplies

Uninterruptible power supplies (UPSs), also known as backup power or emergency power supplies, are used in sectors such as computer, communication (e.g., telephone network), domestic, military, security (e.g., banks and elevator), industry (e.g., power failure critical processes), medicine, emergency and lighting (e.g., airport) [6, 7, 11, 12, 14–16]. An industrial study shows that there is a wide power range for UPSs from below 1 kW (e.g., single computers) to above 100 kW (e.g., industrial production). UPSs for large computer suites can be even in the order of MWs [119]. TPV technology is particularly suitable for the lower power range up to 10 kW. Above this power, other technologies compete strongly with TPV. In the low power range UPSs are likely to operate within buildings and utilise batteries. Indoor operation of a hydrocarbon based TPV UPS would require usually flue gas extraction, which is assumed to be not justifiable below an electricity output of 100 W. It is assumed that the power is usually supplied from the grid and the TPV UPS covers occasional faults. After a power fault the TPV UPS could be refuelled. This operation mode is unlikely to require high TPV system efficiencies and a target range from 5 to 15% has been assumed. Costs and reliability are likely to be more important aspects.

The indicator for the *technology constraints* and the research and development effort has been rated as negative due to the complex overall design. A TPV UPS system design would include an inverter (not required for DC systems), switches, a means to monitor the grid and a buffer battery. Also no UPS system development could be identified in the TPV community.

Competition arises from batteries in the small power range and ICE generators in the large power range, where disadvantages have been discussed previously for portable generators (see Sect. 8.4.2). For UPSs, often secondary batteries immediately provide power for several minutes and bring a standby diesel generator on the load for long-term backup [120]. Flywheels or capacitors in conjunction with heat engine generators may also be utilised as short-term electrical storage options [119]. UPSs using fuel cells are emerging. Fuel cell lifetime is not critical because of the low operation time (only during power fault). Hence, a TPV UPS system is regarded here to have advantages over deployed technologies, but limited advantages over fuel cells in this application.

In general the UPS *market* is large and growing. There are potentially niche markets for reliable, low noise and fuel flexible solid-state TPV UPSs. Some power failure critical applications make use of a redundant UPS, which may also be a market for TPV. Yet no market push from the TPV community could be identified. As a result the market and cost indicator has been rated as balanced. Regarding the local *human impact* a positive rating has been given, since local pollution and security of supply could improve. Also a TPV system could be more user-friendly (e.g., no regular checks as for heat engine generators or batteries) and operate quietly. A minor global human impact can be expected due to the short-term use.

8.4.4 Combustion Application: Remote Power

For future fuel powered *space* missions, it has been pointed out that advances in hydrogen storage need to be made to meet the requirements. TPV systems have been proposed as an alternative to fuel cells for space applications [121]. However, most TPV space research focuses on radioisotope rather than combustion systems (see Sect. 8.2.2).

For *terrestrial* remote power applications, the power density of the generator is usually uncritical, because operation is typically stationary. In the smaller power range the load has been assumed to be typically constant (e.g., telecommunication repeater), whereas in the large power range the load has been assumed to vary (e.g., non-grid connected households). The fuel for a remote TPV combustion system may be available on-site (e.g., gas and oil exploration) or supplied through regular site visits. Similar to thermoelectric on-site generators [17], it can be assumed that TPV generators operate reliably with a long lifetime. Remote generators typically have to operate under hostile environment conditions (e.g., temperature, humidity).

Figure 8.4 in Sec. 8.3.2 summarises potential options for remote power supply. Currently, primary batteries can supply low power and long operation applications (e.g., battery powered radio temperature sensor) and secondary batteries are suitable for higher power and shorter operation applications (e.g., electric wheelchair or model-making). In less accessible applications requiring both high power and long operation, secondary batteries are typically recharged using a renewable source (e.g., solar PV and wind). Less remote applications allow refuelling of heat engine generators. These generators typically utilise a combustion process as a heat source (e.g., thermoelectric, heat engine, Stirling and TPV), except for fuel cells with a direct conversion. Thermoelectric generators can be advantageous in application with the following factors in their favour: fuel is on-site (e.g., gas and oil exploration), limited availability of renewable sources, high reliability requirement or long lifetime need. TPV generators promise higher efficiencies than thermoelectric generators in this niche market. Hence, the replacement of these thermoelectric generators is considered here as one potential remote TPV application and named *unmanned remote power* (Fig. 8.4, No. 2).

Very remote applications (e.g., in polar regions, deep sea or space) may have limited renewable sources and no fuel on-site. Also battery lifetime may be too short and regular refuelling may be unfeasible. In such applications radioisotope sources may be utilised together with a thermoelectric or TPV generator (Fig. 8.4, No. 4, see Sect. 8.2.2).

Another remote system option with long autonomous operation is a solar TPV system with thermal energy storage (Fig. 8.4, No. 1). Such system is not limited by battery lifetime and the (stored) heat can be converted into electricity by direct heat-to-electricity devices (e.g., TPV) or external heat engine generators (e.g., Stirling). Stirling engines are assumed to be suitable around and above the 10 kW power range. For smaller systems it can be assumed that these generators show disadvantages in terms of the complexity compared to TPV. Also the Stirling engine efficiency decreases for the smaller power range (see Sect. 7.2.2). Heat engine generators are readily available for less remote applications (Fig. 8.4, bottom right), where regular maintenance is not critical. In the intermediate power range centred around 1 kW two TPV system configurations have been identified. The first configuration is solar TPV conversion (Fig. 8.4, No. 1), which is discussed in more detail in the Sect. 8.3.2.

The third configuration (Fig. 8.4, No. 3) consists of a renewable generator (e.g., solar PV and wind) combined with a secondary battery and a combustion driven generator (e.g., TE and TPV). This configuration has the advantage, that both the renewable output power and the battery capacity can be scaled down considerably, when compared to the renewable/secondary battery configuration only. In this case, the combustion generator acts as a backup power supply if the renewable source is not available. It follows a detailed discussion of the two identified options, namely unmanned remote power (No. 2) and remote renewable-TPV hybrid system (No. 3).

Applications of *unmanned remote power* include the areas of water supply (e.g., monitoring and pumping), oil/gas exploration and distribution (e.g., cathodic protection, valve operation and data acquisition), stationary telecommunication (e.g., repeater), environmental monitoring (e.g., weather, air quality and scientific measuring) and navigational aids (e.g., aircraft, shipping, road and rail signalling) [4–6, 15, 16, 18]. These applications have typically a constant load. The reliability requirements depend on the specific application. A large share of these applications can be supplied by a hybrid system consisting of a renewable generator and secondary batteries. There are, however, applications supplied by combustion thermoelectric generators and these would be also the niche applications of interest for TPV.

The *power range* has been identified from approximately 1 W to 1 kW using competing technology literature [13, 16, 17]. The thermoelectric generator is the major direct competitor for this small niche market. Hence the minimum efficiency has been assumed with the same value of 3% as current combustion-powered thermoelectric generator (Sect. 7.4.1). Systems with such low efficiency may be used where the fuel is cheaply available on-site, such as for oil/gas exploration and distribution. The near-term target of 10% has been selected as an upper efficiency limit. This value would be about a 3 times improvement over the efficiency of current thermoelectric generators and this may also open new markets.

The indicator for the *technology constraints* and the research and development effort has been assessed as negative. The low efficiency requirements have been regarded positive. However, negative aspects have been identified as dominating. No TPV systems development for this application could be identified and the operation in hostile environments also contributed negatively. Overall the *competition* and benefit has been rated positive compared to deployed and emerging competitors. TPV systems have demonstrated higher efficiency than thermoelectric systems (deployed). Fuel cells have demonstrated higher efficiencies than TPV, but other TPV advantages are believed to outweigh fuel cells. They include fuel flexibility and storage, as well as the potentially higher lifetime of TPV. *Market* and cost issues have been rated balanced. The TPV community identified the unmanned remote power as a niche market [6], but no work towards a system development could be identified. Another constraint is likely to be the lack of funding of this application. Potential advantages are long operating hours and the allowance for premium capital costs in this market. Portable generators and uninterruptible power supplies may be seen as a similar and potentially large market. No major improvements in terms of the *human impact* can be expected by using TPV instead of a thermoelectric generator in this application. Hence, the human impact of unmanned remote generators has been rated as neither better nor worse.

Another interesting configuration is a *remote renewable-TPV hybrid system* (Fig. 8.4 No. 3)A major drawback of renewable sources (e.g., solar PV, wind and hydro) is the fluctuation in output power (e.g., daily and seasonally). Hence, using a combined renewable and secondary battery system for non-grid connected supply requires large sizing of both secondary batteries and renewable sources to meet variation in supply and demand at all times. A possible solution to this is the use of an additional combustion-powered generator. It has been simulated that the size of the PV system can be reduced to a third of that required for an exclusively PV-based system, if only 10% of the annual demand is met by the combustion generator, assuming a constant load and Central Europe climate. It has been pointed out that this redundancy arrangement can also improve the overall reliability [2].

For the combustion TPV generator, a target power range from 100 W to 10 kW has been assumed. The renewable generator would usually be larger than the TPV generator. Below this power range the complexity of this renewable hybrid system can be regarded as a hindrance in terms of the number of components, optimisation of the size of each component and the energy management. Above 10 kW power major competition arises from heat engine generators (e.g., ICE and Stirling). The hybrid remote power applications may be similar as those for the unmanned remote generator discussed previously. In addition to these unmanned applications, larger power applications could include remote housing (e.g., developing world) or larger telecommunications installations. Varying load demand could be met by a secondary battery. It is also feasible that the combustion TPV system would operate with on-site fuel (e.g., biomass). For fuel cell systems biomass operation is not possible in a simple way.

For the hybrid remote power system an efficiency range with a higher upper limit of 5 to 15% has been adopted, compared to the previously discussed unmanned remote application. It is believed that the higher power range makes the number of site visits for refuelling and fuel costs more important issues.

The complexity of this approach has led to a negative rating of the indicator for *technology constraints* and the research and development effort. Although TPV has been suggested for this application [2], no system development could be identified. The *competition* has been rated as advantageous and is generally regarded as similar to the previously discussed application on unmanned remote power. *Market* and cost issues have been rated as balanced and are also regarded similar to the previous application, where differences should be a shorter operation time (negative) and a larger market (positive). Currently, supplementary combustion generators are likely to be ICE based. Hence a TPV system could improve local *human impact* factors (e.g., reduced pollution due to continuous combustion, lower noise and lower maintenance). Potentially primary energy savings are feasible in the developing world using this configuration. Hence, the highest rating for the human impact has been given.

8.4.5 Combustion Application: Transport Sector

At the time of writing, the transport sector accounts for roughly one quarter of the total primary energy consumption in typical industrial countries. One major route to reduce energy consumption in this sector is the improvement of the poor energy efficiency. For example an automobile engine converts typically not more than 30% (Otto engine) to 40% (Diesel engine) from the fuel energy into useful work (maximum brake thermal efficiency) [122]. These values are for optimum load conditions and decrease further for normal operating conditions. *Battery and hybrid electric automobiles* are considered a major technology option in order to improve this propulsion efficiency. TPV generators in the power range from 6 to 10 kW have been examined for series hybrid automobiles in the US [123, 124] and within a project funded by the EU (ΓÇ£The REVΓÇø) [125]. Potentially, a TPV generator could have advantages in terms of noise, fuel flexibility, power density, reliability and maintenance, when compared to internal combustion generators and fuel cells. The major challenge for TPV generators in hybrid vehicles is the high efficiency requirement. For example in the EU project the fuel-to-electric efficiency target has been set to 35% [125]. This value can be considered as a long-term target for TPV conversion and would require some technological progress.

Harvesting and conversion of exhaust gas energy into electricity could also improve propulsion efficiency, since electricity is typically generated from the shaft power of the propulsion engine. Thermoelectric generators are considered as a major option to convert the energy from the heat engine exhaust gas into electricity [18, 126]. A TPV system has been designed to convert exhaust gas from a gas turbine into electricity. Here, problems have been reported to meet the desired radiator temperature of around 1,300°C [127]. Automobile flue gas temperatures vary up to a maximum of around 1,000°C [126]. It can be concluded that exhaust gas power harvesting may be of interest in future, if advances in TPV conversion towards lower radiator temperature systems are made (e.g., micron-gap systems or efficient filter concepts). For this assessment exhaust gas power harvesting has been excluded because temperatures are typically too low.

The two applications considered in detail are *small power propulsions* and *auxiliary power units*. In general, the component and system requirements in the transport sector are challenging. There is a need for reliable operation in hostile environments (e.g., temperature, humidity and vibrations), as well as requirements for low weight and low maintenance.

Small vehicles in the road, air and water sector can have special requirements in terms of low noise, high reliability or low complexity. A TPV generator could meet these needs and is considered competitive in the smaller power range. Examples include the following [3, 5, 14, 128, 129]:

- Land (e.g., electric wheelchair, electric bike/trike, power assisted bicycle, luggage/pallet trolley, electric cart, recreational vehicle, sweeper, scooter, golf cart, forklift, lawnmower, snowmobile, all terrain vehicle, airport vehicle, station car, robot and remotely piloted vehicle).
- Air (e.g., small aeroplane and unmanned aeroplane).
- Water (e.g., small ship, jet-ski and unmanned submarine).

Currently either secondary batteries together with an electric motor or ICEs typically power these vehicles. Heat engines tend to be used in the larger power range and batteries in the smaller power range. In the intermediate power range, assumed here from 100 W to 10 kW, both concepts have their weaknesses. The battery concept tends to have disadvantages in terms of energy density (weight, operation range), recharging (or refuelling), low temperature operation and lifetime. On the other hand, the heat engine concept has drawbacks in terms of noise, maintenance, local pollution and starting reliability. Hence it is assumed that combustion TPV generators together with electric motors for propulsion could have advantages compared to both existing concepts. Optionally a secondary battery may be applied to supply peak propulsion demand and to store recovered brake energy. In general high TPV conversion efficiencies are required, since high efficiencies reduce fuel to be transported and this results in a long operation time. The electric propulsion motor and possibly a battery will add additional weight. Hence, the efficiency target range has been set to the highest range of this assessment (15–20%).

The *technology constraints* and the research and development effort of this application are generally challenging. No TPV system development could be identified specifically designed for small propulsion power and there are high efficiency requirements. Other negative aspects are complex overall design requirement (energy management, battery and power electronics) and the operation in changing environmental conditions. As already discussed TPV should have specific advantages over deployed *competing* technologies (batteries and heat engines). Fuel cells are identified as the major emerging competitor. It has been assumed that TPV advantages (fuel storage and flexibility and high power density) outweigh the high efficiency of fuel cells. The largely diverse *market* should have niche applications allowing for higher costs and large potential markets at lower costs. Negative market aspects are seen to prevail and they include minor TPV community interest, the short operation time and the unlikelihood of funding. Hence a negative rating has been given. The local *human impact* has been seen as positive. Currently reciprocating heat engines are noisy and polluting (TPV uses continuous combustion) and batteries can be regarded as not user-friendly (recharging, limited operation range, safety). On the other hand, large energy saving can be regarded as unfeasible in this niche market application.

In the transport sector there is a large variety of auxiliary (non-propulsion) power consumers for comfort, safety and control functions. Here *auxiliary power units* (APUs) are of interest. The electricity is utilised for applications such as ignition, lights, starters, navigation devices, electric windlasses, alarm system control, fans, heated windows, TV, Radio and by-wire technologies (e.g., fly-by-wire, break-by-wire and steer-by-wire) [3, 6, 12, 14, 116]. The electricity required for these applications may be generated coupled or decoupled from the propulsion engine.

Typically *coupled* systems use a generator and the shaft power of the propulsion heat engine. This can be regarded as a simple configuration with a generator as the major component. Other coupled systems may utilise the exhaust gas heat of the propulsion engine, which is preferable in terms of the overall efficiency but not commonly used for the small power range (e.g., automobile). Exhaust gas conversion with steam turbines in large ships or thermoelectric generators are rare examples of such configuration. It is self-evident that coupled systems generate only power as long as the propulsion engine is operating. Hence propulsion engine idling for electricity generation can be common (e.g., trucks), but this is undesirable in terms of the fuel consumption and noise. Electrical storage systems can only partly overcome idling. The secondary battery concept in the transport sector has disadvantages in terms of the energy density, self-discharge, low temperature performance and constant "key-off" load supply. For example in the automobile, "key-off" loads (e.g., keyless entry, theft alarm and clock) are sufficient to drain the battery in a state-of-the-art car when left parked for long periods (e.g., at an airport for several weeks) [130]. For some applications the security of electricity supply can be another concern. Systems coupled to a propulsion heat engine are susceptible to a propulsion engine failure. For all these reasons electricity generation decoupled (auxiliary) from the propulsion is of interest.

There are several technology options for auxiliary power units (APUs), such as reciprocating engine generators, fuel cells or direct heat-to-electricity devices. Transport applications with a large power requirement may apply for example gas turbines as APUs (e.g., aeroplane, ship). As already mentioned in the previous paragraph, smaller power applications typically rely on a propulsion engine coupled generation (e.g., automobile, trucks). This smaller power range, assumed here up to a power of 10 kW_{el}, has been identified as a potential market for TPV. TPV APUs could be also of interest down to very small powers in the order of watts to supply "key-off" loads in automotive applications [130]. This "key-off" power is small compared to the propulsion power and the operation time may be limited (e.g., to an airport visit), so that the importance of the efficiency can be regarded as secondary. Hence, a typical thermoelectric generator efficiency of 3% has been assumed as a lower limit.

The *technology constraints* and the research and development effort have been assessed as balanced. The MIT together with a car consortium considered TPV APUs [131]. Another positive aspect is seen here in the low efficiency requirements (fuel on-site) for the small powers. The major challenge is seen in the harsh operation environment (e.g., automobile with a wide temperature and humidity range, as well as vibrations). Fuel cells are identified as one major emerging *competitor*. It can be assumed that TPV advantages, and in particular the flexible operation on any propulsion fuel, outweigh the high efficiency of fuel cells. In the small power range there is some competition from deployed niche market thermoelectric generators, but these systems have a low efficiency. Hence, TPV should have advantages over deployed and emerging technologies. *Market and cost* issues have been rated balanced. In general there is a market requirement for reliable APUs. The automotive market is a large potential market [130] and niche markets are also feasible (e.g., TPV generator working in CHP mode for truck idling). On the other hand, public funding is regarded as less likely. In general hybrid vehicles are more likely to be funded due to their energy saving potential. The local *human impact* has been rated as positive, because local pollution, noise, security of supply and user friendliness can be expected to improve. The global human impact was rated negative, since no major primary energy saving can be expected.

8.4.6 Combustion Application: Combined Heat and Power

Combined heat and power (CHP), also known as cogeneration or total energy, is the simultaneous generation of heat and electricity [132, 133]. Mechanical and cooling power is not considered here further, although it may be also defined as CHP. CHP is an attractive energy saving option because of its high overall efficiency (typically 80–90%) defined as the ratio of useful heat and electricity output to the fuel energy input. This efficiency value can be compared with the low efficiency of centralised fossil fuel power stations that release huge amounts of waste heat to the environment. Cost effective operation of CHP systems typically requires a runtime of several thousand hours per year and decisive economic factors are the capital cost of the CHP system, as well as the fuel and electricity prices [134].

The generated heat varies from low-grade (e.g., hot air or water for space heating) to high-grade (e.g., steam for industrial applications). TPV systems are currently limited to low-grade heat generation, because the heat originates from the PV cell cooling. TPV systems operating with cell temperatures up to about 90°C have been reported [135], but a typical cell temperature may be currently around 60°C in CHP mode. Generally, remaining flue gas heat leaving the TPV system using an additional heat exchanger could upgrade this heat [8, 9]. Hence, overall efficiencies higher than the assumed 80% of this assessment and higher heat supply temperatures than 60°C are regarded as feasible in future.

CHP systems tend to be built on-site and matched to the heat load, because transportation of heat over long distances is usually difficult. Short-term time-varying heat demand can be bridged by thermal energy storage systems. Below 100°C water is a simple and effective thermal energy storage media. Water can also act as the heat carrier media (no heat exchanger required), is widely available and has one of the highest sensible heat contents of any liquid around ambient temperatures [136]. In CHP systems the PV cells are usually cooled by water and the sensible heat of the water can be stored in tanks as in some conventional central heating systems. Long-term variation of the demand does usually not allow matching of supply and demand in all cases. This situation can lead to a larger sized and shorter operating CHP system. A known example is the space heating demand with seasonal variations using an oversized system in the summer. The electricity output tends to be less critical compared to the heat output, since electricity can be consumed on-site and surplus electricity can be exported to the grid. Electrical output powers of traditional CHP plants typically range from 100 kW to 100 MW using primary movers such as steam turbines, gas turbines or reciprocating engines combined with a generator [132]. Recently, several emerging technologies allow smaller-sized systems with electrical output powers as small as around 1 kW. The smallest systems have been named micro CHP [134, 137, 138].

Table 8.4 summarises the reported CHP TPV research in literature. The table lists the research and development status, the PV cell material, the thermal

Table 8.4 CHP TPV developments by different institutions

Application	Institution	R & D status	PV cells	Output power	Efficiency target	Source
Micro CHP	JX-Crystals, US WS GmbH, Germany	Partly built	GaSb	1.5 kW_{el} 12.2 kW_{th}	$\eta_{el} = 12\%$ $\eta_{sys} = 80\%$	[78, 110]
Self-powered heating (Midnight Sun®)	JX-Crystals, US	Beta testing of 20 units	GaSb	7.3 $kW_{th\#}$ 0.1 kW_{el}	$\eta_{el} \sim 1$–2%	[24, 139]
District and industrial heating	JX-Crystals, US	Partly built	GaSb	147 $kW_{th\#}$ 20 $kW_{el\#}$	$\eta_{el} \sim 10$–15%	[77, 140]
Self-powered heating	Quantum Group, US	Demonstrated	Si	0.2 kW_{el} 19.2 $kW_{th\#}$	$\eta_{el} \sim 1\%_{\#}$ $\eta_{sys} > 83\%$	[141]
Self-powered heating, Micro CHP	PSI, Hovalwerk AG	Partly built	Si	0.2–1.5 kW_{el} 10–20 kW_{th}	$\eta_{el} = 1$–5%	[85]
Micro CHP	ISE, Freiburg	Partly built	GaSb	0.13 $kW_{el\#}$ 1.1 $kW_{th\#}$	$\eta_{el} = 7\%$ $\eta_{sys} = 60\%$	[142]
Back-up, Micro CHP	ISET, Kassel	Demonstrated	GaSb	60 W_{el} 1.2 $kW_{th\#}$	$\eta_{el} = 4\%$ $\eta_{sys} = 86\%$	[143]
Micro CHP	Dutch Energy Research Found.	N/A	Si	N/A	N/A	[144]
CHP	British Gas Research & Techn. Centre	N/A	N/A	0.3 kW_{el}	N/A	[144]
Micro CHP, District heating	SERC, Sweden	Partly built	InGaAs GaSb	10–1,000 kW_{th}	N/A	[93]

and electrical output power, the efficiency and the literature sources. Efficiency and power values marked with # have been calculated in this work.

In the presented assessment three application groups were identified. These are self-powered heating devices (discussed in Sect. 8.5.2) and two CHP applications with different power levels: micro CHP (small power), district heating and industrial CHP (large power).

Micro CHP systems aim to replace conventional boilers in a dwelling and provide both electricity and heating to a single-family house. To make CHP suitable as a boiler replacement the size of traditional CHP systems needs to be reduced. There are requirements in terms of high reliability, low local pollution, low noise and low costs [134, 145]. Both US and European studies have identified an efficiency target from fuel-to-electricity in the range from 10 to 25% and an overall efficiency target of 80% or higher [134, 145]. The 10% electrical efficiency has been adopted as a minimum target in this assessment. Generally, no deployed technology can fulfil all these requirements. At the moment, there are several technologies under active investigation for micro CHP systems each with their own drawbacks and strengths. At the time of writing, these small sized CHP systems are still in a development and demonstration phase [146].

Reciprocating internal combustion engine (ICE) systems have been traditionally the CHP systems with the smallest output power among the traditional CHP technologies. Adapted reciprocating engine generators are under development to improve noise, local pollution and maintenance. Some systems with 5 kW_{el} output power are commercially available in Europe and smaller systems seem to emerge. Typical thermal efficiencies of small sized ICEs are 60% and electrical efficiencies are around 25% based on the lower calorific value of the fuel [146]. Small *Stirling engine* systems have become commercially available or are being developed. Stirling engines can have the advantage of a cleaner and more fuel flexible combustion compared to reciprocating ICEs. The electrical efficiency drops from large units (~30%, 50 kW) to small micro CHP units (~10%, 1 kW). *Rankine* systems with an electrical power of 1 KW to a few KW are also emerging from research and development laboratories [145]. Traditionally steam turbines have been mainly used for large-scale centralised power generation. For micro CHP, the working fluid water/steam is replaced by an organic fluid (Organic Rankine Cycle ORC). Generally the advantages are similar to Stirling systems (both are external heat engines). Rankine systems for micro CHP are considered to be in a less developed stage compared to Stirling systems and small systems show similar electrical efficiencies (~10%). At the moment, there are two major *fuel cell* types considered for micro CHP. These are SPFC and SOFC (see Sect. 7.3.2) [145]. Both types operate in field trials. The major advantage of fuel cells compared to other micro CHP technologies is the high electrical efficiency (30–40%). The major drawbacks have been identified to be capital cost, overall efficiency (e.g., for SPFC) and lifetime [146, 147]. *AMTEC* micro CHP systems have also been proposed [20]. *Thermoelectric* systems have also been considered for systems with a low electrical efficiency (self-powered operation) [70]. Future advances in thermoelectric system efficiency may also allow micro CHP operation.

For micro CHP-HTPV systems a target *power range* from 1 to 10 kW has been identified from the reviewed competing systems. A typical TPV system may aim for a power output of around 1 kW with stronger competition from other technologies in the power range close to 10 kW [9].

No major *technology constraints* could be identified. There have been developments towards prototypes by more than one institution and system performances could be partly demonstrated. The inverter and grid connection issues have been assessed for solar PV systems previously and have also been demonstrated for TPV operation [148].

There are no fully deployed *competing* technologies for micro CHP, but several emerging technologies as discussed previously. At the moment these are mainly ICE, Stirling, ORC, SPFC and SOFC systems [146]. The long-term strength of TPV is thought to be the solid-state operation. This should allow building micro CHP systems with the same maintenance requirements and reliability performance as existing boilers. The long-term energy efficiency of buildings is likely to improve, so that smaller-sized systems would be required. However, heat engine efficiencies tend to decrease for smaller sizes and scaling to various sizes could be also simpler for a TPV system. Fuel cells are also solid-state devices, have demonstrated high efficiencies and can be flexibly scaled. Nevertheless, even if all current fuel cell problems for micro CHP could be overcome (e.g., lifetime, economics), there could be still a long-term market for oil and biomass based micro CHP-TPV systems, since the fuel flexibility is another challenge of fuel cells. This discussion points out that there are specific TPV benefits over emerging competitors. On the other hand, there are a number of emerging technologies and they are close to the market. This is considered to outweigh specific TPV advantages. Hence, overall TPV has been not rated advantageous over its emerging competitors.

There is a huge mass *market* for micro CHP. For example, Great Britain and Germany have a huge gas boiler market that would be potentially suitable for micro CHP [138]. Important market aspects include a high total dwelling stock with central heating, a extensive natural gas distribution network, high annual boiler sales, liberalised gas and electricity market, suitable heating loads and suitable regulation to facilitate the connection of embedded generation [134]. There are also niche markets for non-grid connected systems and self-powered boilers allowing for higher capital costs. Funding is generally feasible, since the promise of primary energy saving has stimulated political interest (e.g., in the EU). TPV cost estimations indicate that TPV could compete with fuel cells and heat engines [85]. Hence, market and cost issues have been rated as positive.

The *human* impact has been identified as very positive (rating 3). Micro CHP has the potential of considerable primary energy savings, if large numbers of central heating boilers could be replaced [134]. The slight increase in local energy consumption (or local pollution) of a micro CHP system compared to current conventional boilers is not seen as a major disadvantage. In long-term micro CHP systems promise cost savings with a slight increase of fuel costs but major savings in electricity costs for the user. Micro CHP is also a potential technology for grid-independent backup power in case of a power failure.

In addition to micro CHP (1–10 kW) there has also been interest in larger sized TPV systems. TPV systems with thermal powers of over 100 kW and even up to 1,000 kW for apartment, *district and industrial heating* have been proposed [77, 93, 140]. This results in an electrical *power range* from about 10 to 100 kW assuming the same *efficiency* range as for micro CHP (10–15%). Scaling-up of the TPV output power has not been demonstrated yet. However, TPV systems can operate modular as proposed by Fraas et al. [77], so that *technology constraints* similar to those for micro CHP can be expected. *Competition* arises from both deployed and emerging technologies. Deployed reciprocating ICE CHP systems are commercially available. Maintenance and noise of these engines have been regarded as acceptable for the large CHP range compared to micro CHP systems (e.g., installation in a separate room may be common). Small gas turbines, known as micro-turbines, are an emerging technology. The Stirling engine with an output power of tens of kWs is another emerging technology [149]. It is concluded that there is strong competition from both deployed and emerging technologies and no clear overall benefit could be identified using TPV (Rating 1).

Promising niche *markets* are applications with a steady heat demand over the course of a year. Examples are swimming pools, leisure centres or industrial heat processes [150]. There has been some cooperative research of the industry and a TPV company (ABB with JX-Crystals). The feasibility of primary energy saving should allow funding. In terms of capital system costs per output power, it needs to be considered that ICE systems become increasingly cheaper for larger output powers. TPV systems are likely to be mainly determined by PV cell costs. Hence, TPV system costs per output power can be expected relatively constant for increased output powers. One positive aspect is that industrial and distributed heating CHP systems could have longer operation hours to improve economics compared to micro CHP systems. Overall the market and cost indicator has been rated balanced. It has been assumed that there are generally suitable market conditions, but some cost constraints. Similar considerations have been made for the *human impact* as for micro CHP.

8.5 Waste Heat Recovery Generator

8.5.1 Waste Heat Source

Waste heat applications are defined in this work as those where the major purpose is heat rather than electricity generation. In such applications TPV systems can converts a small share of the waste heat into electricity. Applications using waste heat sources have the advantage of the availability of free or low cost heat. In some cases this heat is accessible without interruption, such as for industrial high-temperature processes operating throughout day and night. Finally the efficiency is often not a decisive factor in these applications where heat is generated in any case. Two applications are identified. These are self-powered heating (Sect. 8.5.2) and industrial waste heat recovery (Sect. 8.5.3).

8.5.2 Waste Heat Application: Self-Powered Heating

Self-powered heating is of interest where operation during power faults is critical or no public electricity grid is available. The end-use of the hydrocarbon based combustion heat includes applications such as for space heating, hot water and drying. Examples are self-powered central heating units, district heating units, space heaters with fans, portable heaters, tent heaters, independent vehicle heaters or self-powered furnaces [3, 4, 18, 20, 139]. For example modern central heating units cannot generate heat, if the public electricity grid fails. Hence, a self-powered operation independent of the public electricity grid is of interest. TPV work in the electrical power range from 10 W to 10 kW has been reported. Applications included a combined camping heater and generator [151], an off-grid stove [139], self-powered central heating units [85, 141] and heating units for apartment building [77, 140, 152]. Similarly, critical industrial heating processes may have additional backup power systems in order to ensure the operation during power faults. Also here, self-powered operation of industrial processes using a TPV heat recovery system would be advantageous.

Typically the heating devices have a much larger thermal power compared to the electrical power requirements. The end-use of the electricity is typically for control and heat distribution. Usually heat is distributed by forced convection (e.g., pumps, fans). The smallest thermal power has been assumed to be around 1 kW and this value corresponds roughly to the thermal power of a hairdryer. District heating units with thermal powers up to 1 MW also could have a need for grid-independent operation and have been assumed as an upper power range limit. From different sources [85, 139, 141] it becomes clear that these devices typically have an electricity requirement in the order of 1% of their total heat consumption. Consequently an electrical power range from 10 W to 10 kW has been calculated from the thermal range.

Lower electrical powers than 10 W are also feasible. Work by Goldstein and DeShazer [153–155] and the author [156] demonstrated that radiation could be extracted from a process by thermally stable dielectric solid radiation guides (see Sect. 6.3.4). In this arrangement one side of the dielectric solid light guide is in the high-temperature heat source environment, where radiation is coupled into the guide, and the other end can have a low temperature and illuminates the PV cell. Using total internal reflection allows radiation guidance. For example, thermoelectric generators typically utilise conduction to transfer heat to the thermoelectric modules. Thermal conduction can be associated with temperature gradients. Hence, thermoelectric modules need to be placed close to the heat source in order to achieve high hot side temperatures. On the other hand, an optical guide can transfer radiation with little losses over long distances. It can be assumed that the radiation guide is simpler than other TPV systems using a cavity with challenging high-temperature mirror design. The light guide arrangement could be inserted into any high-temperature process with a suitable temperature. It seems likely that such arrangement could be also used to convert small amounts of (waste) heat into electricity. Feasible applications include self-powered heating devices (e.g., powering of displays or thermostats), self-powered sensors and fire energy converters for camping. Goldstein also proposed the conversion of jet engine heat into electricity using dielectric solid light guides [154].

The variety of potential applications and technological options results in a wide *power range*, which has been assumed from mW to 10 kW. Similarly, efficiency requirements in a wide range can be expected. A marginal *efficiency* would be sufficient to power a small sensor (e.g., 10 mW electrical from 10 W thermal). On the other hand, a self-power high-temperature process may have high efficiency requirements. For example a high-temperature process with a power of 1 MW_{th} could require 1% of its thermal input as electricity (10 kW_{el}). Assuming a TPV efficiency $\eta_{TPV} = 13\%$ results in 67 kW_{th} low-grade heat. Such low-grade heat may or may not be utilised. Such generator would extract 77 kW_{th} or about 8% of the total thermal power as heat. The example shows that, unless there is a need of low-grade heat, the efficiency of the TPV generator for self-powered operation of high-temperature processes should be high. Nevertheless the extraction of 8% of the total heat input seems feasible and the maximum efficiency target was set to the near-term value of 13%. If low grade heat can be utilised, such as for space heating, the TPV efficiency requirements can be much lower. For example, a self-powered central heating unit with 20 kW_{th} could deliver 10 kW_{th} to the TPV system. The TPV system could convert 2% of this heat into 200 W_{el} or 1% of the total heat input. The 9.8 kW_{th} low-grade heat from the PV cells may either be used directly or upgraded by the remaining 10 kW_{th}.

The *technology constraints* and the research and development effort have been rated as positive. Positive factors were the (partly) low efficiency requirements (partly) simple overall design (e.g., sensor powered by light guides) and beta tests of the Midnight Sun® oven [139]. *Competing* deployed technologies are not available in this niche market. There is some experience with self-powered heating systems using thermoelectric generators [157]. Thermoelectric generators are limited in terms of efficiency. It is currently challenging to design a thermoelectric generator with more than 3% efficiency [17]. Hence, TPV conversion is seen superior over its competitors in this configuration. The *market and cost* indicator has been rated as positive. Self-powered heating devices are used in niche markets, where higher system capital costs could be justified. In long-term this application could grow into the large micro CHP and industrial waste heat recovery market. The US Gas Research Institute identified self-powered devices as the major future market for TPV [4] and there is general interest in the TPV community for this application (e.g., Midnight Sun® system). Hence, market and cost issues have been rated as positive. TPV could improve security of electricity supply during power faults. This resulted in a positive local *human impact* rating. The total energy consumption is largely unaffected for the small share of electricity generated (negative global human impact factor).

8.5.3 Waste Heat Application: Industrial High-Temperature Processes

In the large power range, the recovery of *industrial high-temperature waste heat* by a TPV system has been proposed by Coutts [158]. One major attraction of industrial waste heat recovery is that the heat is freely and often steadily available and that the electricity generated can be typically used at the industrial site. The low-grade heat generated by the TPV system may also be utilised for space heating if a demand can be identified. The potential industries for high-temperature waste heat recovery include the iron and steel, non-ferrous, bricks, refractories, cement, ceramics and glass sectors. In particular the glass industry has been suggested [140, 152]. An US company (JX-Crystals) received funding from the US Department of Energy and the company partly developed TPV generator tubes each with an output power of 5 kW_{el} [77, 140, 152]. These tubes could be inserted in high-temperature processes in order to generate electricity and low-grade heat in the form of hot water. Two applications have been proposed, which are high-temperature waste heat recovery in the manufacturing industry and CHP supply for apartment buildings [77, 152]. Yamaguchi et al. [8, 9] assessed the potential of waste heat recovery in Japan. They found that waste heat recovery from flue gas for large industrial processes is already utilised for large furnaces using steam turbines based systems and hence they concentrated on small furnaces where they found temperatures to be too low for TPV operation. However, it was pointed out, that it is possible to upgrade small-scale flue gas heat using premium heat produced by an additional combustion process. Heat recovery from other locations than the flue gas was not considered in their work [8, 9]. The potential of TPV for heat recovery in the high-temperature industry in the UK was assessed by the author [159]. It was found that an overall assessment of TPV in those industries is complex mainly because of the large process diversity and because an individual assessment for each process in terms of TPV use is required. Hence, one example process for each of the three principal locations for heat losses has been assessed in detail. The principal locations and specific processes were product heat recovery on a continuous curved caster, flue gas heat recovery on a regenerative glass tank furnace and wall heat recovery on a 3-phase alternating current electric arc furnace.

Nearly all of the high-temperature processes coincide with a general type of process (Fig. 8.5). The energy entering the process chamber derives from electrical heating, combustion, a hot feedstock, or any combination of the previous sources. Within the process chamber the temperature exceeds 1,000°C. The combustion and feedstock/product may sometimes show interactions. Examples are the

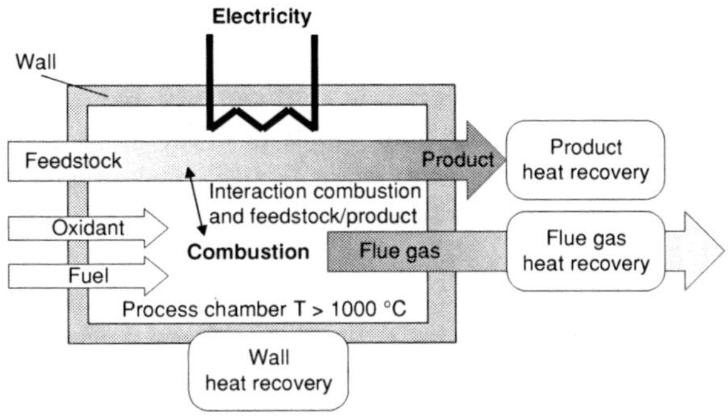

Fig. 8.5 Schematic of a general type of industrial high-temperature process

formation of CO_2 from glass forming reactions and the use of coke in a blast furnace as feedstock and fuel. The feedstock can be charged in different modes including batch, tap and charge or continuous type. The feedstock can consist of raw materials, recycled products or a mixture of both. Typically the specific energy consumption (SEC) of processes using recycled products is lower if compared with raw material processes. The important parameters to specify an industrial high-temperature process are the process temperature and pressure, fuel requirements, insulation losses and thermal efficiency, as well as the flow of mass and energy in the process [160]. The energy flow can be visualised in a Sankey diagram. Hence, this diagram can be used to identify major heat losses for possible heat recovery locations.

For the general type of high-temperature process considered (Fig. 8.5), there are three principal locations for heat recovery, namely product (1), flue gas (2) and wall heat recovery (3). Existing heat recovery methods most commonly utilise the flue gas location (e.g., glass furnace regenerators, pig iron blast furnaces) followed by the product and by-product (e.g., cement clinker cooler, blast furnace slag recovery [161]).

TPV systems could also utilise the sensible and latent heat of products and by-products. Other recovery technologies already utilise the calorific value of by-products. Examples are coke oven, blast furnace and basic oxygen furnace gases in the steel industry [162]. Insulation heat losses are rarely recovered and a rare example is the recovery of radiation losses from a shell of a cement rotary kiln [162].

Once potential sources of waste heat are identified, the end-use of the recovered energy needs to be considered. The use of the recovered energy in the same process (e.g., combustion air pre-heating, feedstock preheating) has several advantages, such as minimising the process SEC, avoiding long distance transport of heat and usually temporal matching of energy supply and demand [163]. However, the use of waste heat within the process is not always feasible and therefore this heat is available for external use (e.g., space heating, electricity generation). External use of energy may be classified by the type of energy conversion: heat-to-heat (e.g., steam or hot water production), heat-to-chemical energy (e.g., forming methane from blast furnace slag) or heat-to-electricity. Heat-to-heat conversion, mostly associated with heat exchangers has simplicity, but is limited to applications where heat recovery and demand coincide since heat can only be transported to a limited extent. Additionally, heat exchangers may have problems with leaks, low efficiencies and fouling from flue gas, especially for temperatures higher than 1,000°C [163, 164]. Heat recovery technologies that generate electricity offer more flexibility since electricity can be transmitted with lower losses than for heat and electricity can be converted to other energy forms at high conversion efficiencies.

For thermoelectric systems it has been pointed out that a low *efficiency* is not a serious drawback in the conversion of free waste heat and that the capital cost per watt is the decisive economic factor [164]. These considerations can also be brought forward for TPV and thus the lower efficiency target range has been assumed with the demonstrated efficiency (6% for non-combustion systems).

The use of 200 tubes each with 5 kW has been suggested in a glass furnace [152]. This would result in an electricity output *power* of 1 MW. This value has been assumed as an upper power range limit for this assessment. Small high-temperature furnaces could also have a requirement for self-powered operation, which would result in a minimum electrical power of 1 kW assuming 1% of a 100 kW thermal furnace.

A partly demonstrated TPV system indicates that there should be no major *technology constraints* [77]. Also a TPV heat recovery system does not require the design of an efficient combustion unit. This should result in a simpler system design. The integration of a TPV system in an industrial process requires some further consideration (e.g., adaptation to process temperature levels and temporal variations). Hence, a balanced rating has been given for the technology constraints and research and development effort.

Competing technologies are external heat engines and direct heat-to-electricity converters. External heat engine based systems operate with different thermodynamic cycles, such as the Rankine or Stirling cycle. These cycles can use different working fluids and may be of open or closed type. This results in a variety of potential recovery systems using external heat engines [165, 166]. For high-temperature waste heat sources of interest here (above 1,000°C), the inlet temperature at the heat engine is usually lower than 1,000°C. There may be different reasons for this including high-temperature engineering difficulties and deliberate cooling for the conversion (e.g., because the heat engine operates at a lower inlet temperature). Also, typically the heat needs to be piped to the heat engine, which inevitably leads to some collection and piping heat losses with heat degradation. One deployed competing combination of technologies generates electricity from heat in the flue gas. This combination consists of a heat exchanger, a high-pressure steam boiler, a condensing steam turbine and a generator [163, 165]. However, the complexity of this approach associated with high capital and maintenance costs limits applications to large industrial plants using only the flue gas heat. On the other hand, TPV systems could be scaled in a wide power range. Also, TPV systems could not only recover flue gas heat, but such system could also convert heat of hot products leaving the process and heat lost through walls. Emerging competing technologies include heat engine concepts based on the organic

Rankine [166], the Stirling [167] and the air-bottoming cycle [165]. Heat engines are usually limited to the conversion of flue gas heat. Product and wall heat losses are typically not converted into electricity using external heat engines.

Direct heat-to-electricity conversion devices have advantages in terms of low maintenance, high reliability, good scalability and low complexity compared to heat engines. In addition, they could be used directly at any heat source (flue gas, wall, product) so that collection and piping losses could be avoided. The major direct heat-to-electricity conversion devices currently of research interest are thermoelectric generators [164, 168]. However, these generators can commonly convert only low-grade heat with hot side temperatures of around 400 K or lower temperatures [164, 168, 169]. High-temperature thermoelectric generators are feasible but there are still engineering challenges to be overcome (Sect. 7.4.1). For high-temperature thermoelectric generators, sufficient and reliable electrical contacts and mechanical properties are difficult to achieve [169]. Hence, thermoelectrics could recover low-temperature heat (<400 K) and TPV the high-temperature heat (>1,300 K). Advances in both technologies may allow the entire temperature range to be covered using direct energy conversion devices. Hence, it can be argued that thermoelectric and TPV generators are not competing with each other, but they should be seen as complementary.

For example, classified by type of use, high-temperature processes accounted for about one quarter of the industrial energy consumption in the UK in 1999. High-temperature processes can be further classified by sector. Most of these sectors have energy-intensive processes operating at temperatures above 1,300 K and which are suitable for TPV operation [160]. This shows that there is a large *market* for industrial waste heat recovery. Niche markets could be self-powered furnaces and backup power supplies in the industry. Funding has been available in the US [152] and very long operation hours should keep payback periods low. From these considerations a positive market and cost rating has been given. The local and global *human impact* has been rated as positive. Primary energy savings are feasible and no additional pollution due to TPV operation should occur. Electricity generation by TPV would be expected to require low maintenance. On-site use of electricity is another environmental advantage.

8.6 Summary

The assessment has been an iterative process, which considered the availability of different heat sources, the capabilities of TPV and competing conversion technologies, as well as the application requirements. The applications have been classified by heat source and CHP mode. Table 8.5 gives the final result of the assessment. Figure of merit ranges for both efficiency and electrical power output of each application have been identified. The four indicator groups have been rated from 1 to 3. The applications with the highest potential were identified by summing these indicators for each application (Table 8.5). It is in the nature of such

Table 8.5 Potential TPV application classified by heat source and CHP mode. For each application figures of merit for the electrical efficiency and power range have been identified. The four indicators are summed up to the final result

	Figure of merit ranges		R & D Effort	Competit. & benefit	Market & cost	Human impact	Final result
	Electrical efficiency (%)	Electrical power					
Combustion non-CHP							
Portable generator	5–15	1 W to 1 kW	2	3	3	2	10
Uninterruptible power supply	5–15	100 W to 10 kW	1	2	2	2	9
Unmanned remote power	3–15	1 W to 1 kW	1	2	2	1	6
Remote renewable hybrid system	5–15	100 W to 10 kW	1	2	2	3	8
Small vehicle propulsion	15–20	100 W to 10 kW	1	3	1	2	7
Vehicle auxiliary power units	2–15	1 W to 10 kW	2	3	2	2	9
Combustion—CHP							
Micro CHP	10–15	1 kW to 10 kW	3	2	3	3	11
District/industrial CHP	10–15	10 kW to 100 kW	3	1	2	3	9
Solar							
Solar TPV storage/combustion	13–25	100 W to 10 kW	1	3	3	3	10
Nuclear							
Space radioisotope generator	19–25	10 W to 1 kW	1	3	3	1	8
Waste heat							
Self-powered heating devices	<13	1 mW to 10 kW	3	3	3	2	11
Industrial waste heat recovery	6–19	1 kW to 1 MW	2	3	3	3	11

assessments that the assumptions have a decisive impact on the results. In general, it can be pointed out that TPV systems could be potentially utilised in all of the twelve identified application fields. The human impact considers the following factors: primary energy savings, low NO_x, low noise, improvement of the security of supply and user friendliness.

Hydrocarbon fuels are widely available, can be easily stored, transported or refuelled, and have high energy density. Combining fuel properties and TPV technology capabilities results in several potential applications. Competing technologies for these applications are mostly batteries in the smaller power range and internal combustion generators in the larger power range. Here, fuel cells are seen as the major future competitor. Portable power applications have been identified as most promising in this group. Military interest and another TPV assessment for Japan also indicate the potential. APUs operating independently of the vehicle propulsion have had less attention in the TPV community, but are regarded here as an application field with a high potential. In particular, applications with low noise, high reliability or low complexity requirements should hold promise.

In CHP systems both heat and the electricity output are utilised. Micro CHP systems aim to replace conventional boilers in a dwelling and have been identified as one of the most suitable TPV application of this assessment. In Western Europe and the US there is some major interest in micro CHP. Generally TPV technology capabilities match the micro CHP requirements, namely high reliability, low maintenance and low noise. Currently, there are several other emerging technologies with similar capabilities operating in field trials and this competition can be regarded as one major challenge for TPV developments. Industrial and distributed CHP systems could have longer operation hours to improve economics and have also been identified as a promising.

At the current stage, *solar* concentrator PV systems outperforms solar TPV systems in terms of efficiency. It is concluded here that even if solar TPV systems could not demonstrate their potentially high efficiency, such systems would have still some unique advantages. Solar TPV systems can use a thermal storage or additional hydrocarbon combustion to bridge solar fluctuations and ensure a flexible and reliable supply of electricity. CHP operation and biomass utilisation of such systems is also feasible. In particular smaller autonomous operating systems without an electrical grid could be of interest.

Space power for near sun and deep space missions is a niche market for *radioisotope* generators. TPV generators could meet the requirements of this high value niche market with unique advantages in terms of lifetime, power density and efficiency compared to other technologies (e.g., TE, Stirling and AMTEC).

Self-powered heating can be considered as an ideal near-term market. Particularly suitable applications are those with a requirement of low-grade heat such as for space heating. This results in low efficiency requirements that have been demonstrated. In long-term, this niche market could grow in larger primary energy saving markets such as industrial waste heat recovery and micro CHP. TPV *high-temperature industrial waste heat recovery* has been also identified as a very suitable application. Potential advantages are primary energy savings, high operation time, low competition from other technologies and moderate efficiency requirements due to the free or low cost of the thermal input. Also, many high-temperature processes are susceptible to power failures, where TPV could act as a backup power source. This high value backup market could be used to launch the TPV heat recovery systems.

References

1. Doyle EF, Becker FE, Shukla KC, Fraas LM (1999) Design of a thermophotovoltaic battery substitute. 4th NREL conference on thermophotovoltaic generation of electricity, Denver, Colorado, 11–14 Oct 1998. American Institute of Physics, pp. 351–361
2. Steinhüser A, Hille G, Kügele R, Roth W, Schulz W (1999) Photovoltaic-Hybrid Power Supply for Radio Network Components. Proceeding of the Intelec'99, Kopenhagen, 6–9. Juni
3. Ralph EL, FitzGerald MC (1995) Systems/marketing challenges for TPV. Proceeding of the 1st NREL conference on thermophotovoltaic generation of electricity, Copper Mountain, Colorado, US, 24–28 July 1994. American Institute of Physics, pp 315–321
4. Krist K (1995) GRI research on thermophotovoltaics. Proceeding of the 1st NREL conference on thermophotovoltaic generation of electricity, Copper Mountain, Colorado, 24–28 July 1994. American Institute of Physics, pp 54–63
5. Rose MF (1996) Competing technologies for thermophotovoltaic. Proceeding of the 2nd NREL conference on thermophotovoltaic generation of electricity, Colorado Springs, 16–20 July 1995. American Institute of Physics, pp 213–220
6. Ostrowski LJ, Pernisz UC, Fraas LM (1996) Thermophotovoltaic energy conversion: technology and market potential. Proceeding of the 2nd NREL conference on thermophotovoltaic generation of electricity, Colorado Springs, 16–20. July 1995, American Institute of Physics, pp 251–260
7. Johnson S (1997) TPV Market Review. Proceeding of the 3rd NREL conference on thermophotovoltaic generation of electricity, Denver, Colorado, 18–21 May 1997. American Institute of Physics, pp xxv–xxvii
8. Yamaguchi H, Yamaguchi M (1999) Thermophotovoltaic potential applications for civilian and industrial use in Japan. Proceeding of the 4th NREL conference on thermophotovoltaic generation of electricity, Denver, Colorado, 11–14 Oct 1998. American Institute of Physics, pp 17–29
9. Yugami H, Sasa H, Yamaguchi M (2003) Thermophotovoltaic systems for civilian and industrial applications in Japan. Semicond Sci Technol 18:239–246
10. Bard J (2005) Thermophotovoltaics—applications and markets, Presentation, 1st conference for thermophotovoltaics: SCIENCE to buisness, 27 Jan, Berlin
11. Honda generators (2010) [Online] Available at: http://www.honda-uk.com/. Accessed 5 May 2010
12. Power Solutions (2010) Kohler Power Systems, US, [Online] Available at: http://www.kohlerpower.com/. Accessed 14 May 2011
13. West C (1986) Principles and applications of stirling engines. Van Nostrand Reinhold, New York
14. List of contents (2009) Batteries to 2012 —market research, market share, market size, sales, demand forecast, market leaders,company profiles, industry trends, Freedonia Group Inc., US [Online] Available at: http://www.freedoniagroup.com/Batteries.html. Accessed 8 May 2010
15. Williams MC (2000) Fuel cell handbook, 5th edn. U.S. Department of Energy, Office of Fossil Energy, National Energy Technology Laboratory, DE-AM26-99FT1675

16. Weston M, Matcham J (2002) Portable power applications of fuel cells, report. Department of Trade and Industry, UK, ETSU F/03/00253/00/REP
17. Thermoelectric generator (2010) Global Thermoelectric, Canada [Online] Available at: http://www.globalte.com/. Accessed 8 May 2010
18. Technical Papers (2010) Hi-Z Technology Inc., US [Online] Available at: http://www.hi-z.com/. Accessed 8 May 2010
19. Overview of the technology (2001) In: Thermionics Quo Vadis, an assessment of the DTRAs advanced thermionics researchand development program, Chap. 3. National Academy Press, pp 15–32, [Online] Available at: http://books.nap.edu/ Accessed 8 May 2010
20. Lodhi MAK, Vijayaraghavan P, Daloglu A (2001) An overview of advanced space/terrestrial power generation device: AMTEC. J Power Sources 103(1):25–33
21. Solar Applications (2002) BP Solar International, US [Online] Available at: http://www.bpsolar.com/. Accessed 22 Sep 2004
22. Volz W (2001) Entwicklung und Aufbau eines thermophotovoltaischen Energiewandlers (in German), Doctoral thesis, Universität Gesamthochschule Kassel, Institut für Solare Energieversorgungstechnik (ISET)
23. Horne E (2002) Hybrid thermophotovoltaic power systems. EDTEK, Inc., US, Consultant Report, P500-02-048F
24. Fraas LM, Avery JE, Huang HX, Martinelli RU (2003) Thermophotovoltaic system configurations and spectral control. Semicond Sci Technol 18:165–173
25. Regan TM, Martin JG, Riccobono J (1995) TPV conversion of nuclear energy for space applications. Proceeding of the 1st NREL conference on thermophotovoltaic generation of electricity. Copper Mountain, Colorado, 24–28 July 1994. American Institute of Physics, pp 322–330
26. Vicente FA, Kelly CE, Loughin S (1996) Thermophotovoltaic (TPV) applications to space power generation. Proceeding of the 31st intersociety energy conversion engineering conference, IEEE, pp 635–640
27. Schock A, Kumar V (1995) Radioisotope thermophotovoltaic system design and its application to an illustrative space mission. Proceeding of the 1st NREL conference on thermophotovoltaic generation of electricity. Copper Mountain, Colorado, 24–28 July 1994. American Institute of Physics, pp 139–152
28. Schock A, Or C, Mukunda M (1996) Effect of expanded integration limits and measured infrared filter improvements on performance of RTPV system. Proceeding of the 2nd NREL conference on thermophotovoltaic generation of electricity, Colorado Springs, 16–20 July 1995. American Institute of Physics, pp 55–80
29. Day AC, Horne WE, Morgan MD (1990) Application of the GaSb solar cell in isotope-heated power systems. Proceeding of the 21st IEEE photovoltaic specialists conference, IEEE, pp 1320–1325
30. Schock A, Mukunda M, Or C, Summers G (1995) Analysis, optimization, and assessment of radioisotope thermophotovoltaic system design for an illustrative space mission. Proceeding of the 1st NREL conference on thermophotovoltaic generation of electricity, Copper Mountain, Colorado, 24–28 July 1994. American Institute of Physics, pp 331–356
31. Schock A, Or C, Kumar V (1996) Small radioisotope thermophotovoltaic (RTPV) generators. Proceeding of the 2nd NREL conference on thermophotovoltaic generation of electricity, Colorado Springs, 16–20 July 1995. American Institute of Physics, pp 81–97
32. Schock A, Or C, Kumar V (1997) Design and integration of small RTPV generators with new millennium spacecraft for outer solar system. Acta Astronautica 41(12):801–816
33. Murray CS, Crowley CJ, Murray S, Elkouh NA, Hill RW, Chubb DE (2004) Thermophotovoltaic converter design for radioisotope power systems. Proceeding of the 6th international conference on thermophotovoltaic generation of electricity, Freiburg, Germany, 14–16 June 2004. American Institute of Physics, pp 123–132

34. Cassedy ES, Grossman PZ (1998) Nuclear fission technology, Chap 7; and The nuclear fuel cycle, Chap 8. In: Cassedy ES, Grossman PZ (eds) Introduction to energy—resources, technology, and society, 2nd edn edn. Cambridge University Press, Cambridge, pp 169–229
35. Bryan K, Smith MS (2003) Definition, expansion and screening of architectures for planetary exploration class nuclear electric propulsion and power systems, MSc thesis, Massachusetts Institute of Technology
36. Foster AR, Wright RL Jr (1983) Radioisotope application. In: Foster AR, Wright RL (eds) Basic Nuclear Engineering, Chap 7. Allyn and Bacon, Newton, pp 168–194
37. Kulcinski G (2001) Lecture 21-nuclear power in space [Online] Available at: http://silver.neep.wisc.edu/neep533/FALL2001/lecture21.pdf. Accessed 28 April 2010
38. Aldo da R (2009) Fundamentals of renewable energy processes. Elsevier Science, Amsterdam
39. Massie LD (1991) Future trends in space power technology. IEEE Aerospace Electron Syst Mag 6(11):8–13
40. Adams LJ (1994) Spacecraft electric power. In: Adams LJ (ed) Technology for small spacecraft, Chap 4. National Academy Press, Washington
41. Wilt D, Chubb D, Wolford D, Magari P, Crowley C (2007) Thermophotovoltaics for space power applications. Proceeding of the 7th world conference on thermophotovoltaic generation of electricity, Madrid, 25–27 Sept 2006. American Institute of Physics, pp 335–345
42. Würfel P (1995) Physik der Solarzellen (in German), 2nd edn. Spektrum Akademischer Verlag, Heidelberg
43. Meschede D (2002) Gerthsen Physik (in German), 21st edn. Springer, Heidelberg
44. AM0 ASTM E-490-00 (2004) Renewable Resource Data Center, supported by the National Center for Photovoltaics at theNational Renewable Energy Laboratory, US [Online] Available at: http://rredc.nrel.gov/solar/spectra/. Accessed 28 April 2010
45. AM1.5 ASTM G33-03 (2004) Renewable resource data center, supported by the National Center for Photovoltaics at theNational Renewable Energy Laboratory, US [Online] Available at: http://rredc.nrel.gov/solar/spectra/. Accessed 28 April 2010
46. Dang A (1986) Concentrators: A review. Energy Convers Manag 26(1):11–26
47. Nitsch J (2008) Lead study 2008—Further development of the "Strategy to increase the use of renewable energies" within the context of the current climate protection goals of Germany and Europe, Report, German Federal Ministry for the Environment (BMU)
48. Winston R, Cooke D, Gleckman P, Krebs H, OGallagher J, Sagie D (1990) Sunlight brighter than the sun. Nature 346:802
49. Gordon JM (2007) Concentrator optics. In: Luque A, Andreev V (eds) Concentrator Photovoltaics, Chap 6. Springer, Heidelberg, pp 113–132
50. Luque A (2007) Solar thermophotovoltaics: combining solar thermal and photovoltaics. Proceeding of the 7th World conference on thermophotovoltaic generation of electricity, Madrid, 25–27 Sept 2006. American Institute of Physics, pp 3–16
51. Swanson RM (1979) A proposed thermophotovoltaic solar energy conversion system. Proc. IEEE 67(3):446–447
52. Stone KW, Leingang EF, Kusek SM, Drubka RE, Fay TD (1994) On-Sun test results of McDonnell Douglas' prototype solar thermophotovoltaic power system. Proceeding of the 24th IEEE Photovoltaic Specialists Conference, IEEE, pp 2010–2013
53. Stone KW, Kusek SM, Drubka RE, Fay TD (1995) Analysis of solar thermophotovoltaic test data from experiments performed at McDonnell Douglas. Proceeding of the 1st NREL conference on thermophotovoltaic generation of electricity, Copper Mountain, Colorado, 24–28, July 1994. American Institute of Physics, pp 153–162
54. Stone K, McLellan S (1996) Utility market and requirements for solar thermophotovoltaic system. Proceeding of the 2nd NREL conference on Thermophotovoltaic Generation of Electricity, Colorado Springs, 16–20 July 1995. American Institute of Physics, pp 238–250
55. Fatemi NS (1996) A solar thermophotovoltaic electrical generator for remote power applications, Final Report, Essential Research Inc., Cleveland, US, NAS3-27779

56. (2004) Thermophotovoltaic Research, Presentation, Space Power Institute, Auburn University, [Online]. Accessed 10 May 2004
57. Chubb DL, Good BS, Lowe RA (1996) Solar thermophotovoltaic (STPV) system with thermal energy storage. Proceeding of the 2nd NREL conference on thermophotovoltaic generation of electricity, Colorado Springs, 16–20 July 1995. American Institute of Physics, pp 181–198
58. Yugami H, Sai H, Nakamura K, Nakagawa N, Ohtsubo H (2000) Solar thermophotovoltaic using Al2O3/Er3Al5O12 eutectic composite selective emitter. Proceeding of the 28th IEEE photovoltaic specialists Conference, IEEE, pp 1214–1217
59. Guazzoni GE, Rose MF (1996) Extended use of photovoltaic solar panels. Proceeding of the 2nd NREL conference on thermophotovoltaic generation of electricity, Colorado Springs, 16–20 July 1995. American Institute of Physics, pp 162–236
60. Demichelis F, Minetti-Mezzetti E (1980) A solar thermophotovoltaic converter. Solar Cells 1(4):395–403
61. Davies PA, Luque A (1994) Solar thermophotovoltaics: Brief review and a new look. Sol Energy Mater Sol Cells 33:11–22
62. Stone KW, Chubb DL, Wilt DM, Wanlass MW (1996) Testing and modelling of a solar thermophotovoltaic power system. Proceeding of the 2nd NREL conference on thermophotovoltaic generation of electricity, Colorado Springs, 16–20 July 1995. American Institute of Physics, pp 199–209
63. Stone KW, Fatemi NS, Garverick LM (1996) Operation and component testing of a solar thermophotovoltaic power system. Proceedings of the 25th IEEE photovoltaic specialists conference, IEEE, pp 1421–1424
64. Imenes AG, Mills DR (2004) Spectral beam splitting technology for increased conversion efficiency in solar concentrating systems: a review. Sol Energy Mater Sol Cells 84:19–69
65. Fraas L, Avery J, Malfa E, Venturino M (2002) TPV tube generators for apartment building and industrial furnace applications. Proceeding of the 5th conference on thermophotovoltaic generation of electricity. Rome, 16–19 Sept 2002. American Institute of Physics, pp 38–48
66. Sarraf DB, Mayer TS (1996) Design of a TPV Generator with a durable selective emitter and spectrally matched PV cells. Proceeding of the 2nd NREL conference on thermophotovoltaic generation of electricity, Colorado Springs, 16–20 July 1995. American Institute of Physics, pp 98–108
67. Mills D (2004) Advances in solar thermal electricity technology. Sol Energy 76:19–31
68. Green MA, Emery K, Hishikawa Y, Warta W (2009) Solar cell efficiency tables (Version 33) short communication, progress in photovoltaics. Res Appl 3:85–94
69. Andreev VM, Khvostikov VP, Khvostikova OA, Rumyantsev VD, Gazarjan PY, Vlasov AS (2004) Solar thermophotovoltaic converters: efficiency potentialities. Proceeding of the 6th international conference on thermophotovoltaic generation of electricity, Freiburg, Germany, 14–16 June 2004. American Institute of Physics, pp 96–104
70. Butcher T (2003) MicroCHP the next level in efficiency, presentation, national oilheat research alliance technology symposium, New England fuel institute convention, Boston, 9 June 2003 [Online] http://www.bnl.gov/est/files/pdf/No_03.pdf. Accessed 28 April 2010
71. Laing D, Pålsson M (2002) Hybrid dish/stirling systems: combustor and heat pipe receiver development. J Sol Energy Eng 124:176–181
72. Nielsen OM, Arana LR, Baertsch CD, Jensen KF, Schmidt MA (2003) A Thermophotovoltaic Micro-Generator For Portable Power Applications. Proceedings of the 12th international conference on solid state sensors, actuators and microsystems, Boston, June 8–12 pp 714–73
73. Yang WM, Chou SK, Shu C, Xue H, Li ZW (2004) Development of a prototype micro-thermophotovoltaic power generator. J Phys D Appl Phys 37:1017–1020
74. Yang W, Chou S, Shu C, Xue H, Li Z (2004) Effect of wall thickness of micro-combustor on the performance of micro-thermophotovoltaic power generators. Sens Actuators, A 119(2):441–445

75. Yang WM, Chou SK, Shu C, Li ZW, Xue H (2003) Research on micro-thermophotovoltaic power generators. Sol Energy Mater Sol Cells 80:95–104
76. Williams DJ, Fraas LM (1996) Demonstration of a candle powered radio using GaSb thermophotovoltaic cells. Proceeding of the 2nd NREL conference on Thermophotovoltaic Generation of Electricity, Colorado Springs, 16–20 July 1995. American Institute of Physics, pp 134–137
77. Broman L, Marks J (1994) Development of a TPV converter for co-generation of electricity and heat from combustion of wood powder. Proceeding of the 24th IEEE photovoltaic specialists conference, IEEE, pp 1764–1766
78. Fraas L, Avery J, Malfa E, Wuenning JG, Kovacik G, Astle C (2003) Thermophotovoltaics for combined heat and power using low NOx gas fired radiant tube burners. Proceeding of the 5th conference on thermophotovoltaic generation of electricity, Rome, 16–19 Sept 2002. American Institute of Physics, pp 61–70
79. DeBellis CL, Scotto MV, Scoles SW, Fraas L (1997) Conceptual design of 500 watt portable thermophotovoltaic power supply using JP-8 fuel. Proceeding of the 3rd NREL conference on thermophotovoltaic generation of electricity, Denver, Colorado, 18–21 May 1997. American Institute of Physics, pp 355–367
80. Hanamura K, Kumano T (2003) Thermophotovoltaic power generation by super-adiabatic combustion in porous quartz glass. Proceeding of the 5th conference on thermophotovoltaic generation of electricity, Rome, 16–19 Sept 2002. American Institute of Physics, pp 111–120
81. Hanamura K, Kumano T (2004) TPV power generation system using super-adiabatic combustion in porous quartz glass. Proceeding of the 6th international conference on thermophotovoltaic generation of electricity, Freiburg, Germany, 14–16 June 2004. American Institute of Physics, pp 88–95
82. Kumano T, Hanamura K (2004) Spectral control of transmission of diffuse irradiation using piled AR coated quartz glass filters. Proceeding of the 6th international conference on thermophotovoltaic generation of electricity, Freiburg, Germany, 14–16. June 2004. American Institute of Physics, pp 230–236
83. Horne WE, Morgan MD, Sundaram VS, Butcher T (2003) 500 watt diesel fueled TPV portable power supply. Proceeding of the 5th conference on thermophotovoltaic generation of electricity, Rome, Italy, 16–19 Sept 2002. American Institute of Physics, pp 91–100
84. Weinberg F (1996) Heat recirculating burners: principles and some recent developments. Combust Sci Technol 121:3–22
85. Palfinger G, Bitnar B, Durisch W, Mayor J-C, Grützmacher D, Gobrecht J (2003) Cost estimate of electricity produced by TPV. Semicond Sci Technol 18:254–261
86. Mattarolo G (2005) High temperature recuperative burner, Presentation 1. Conference for thermophotovoltaics: science to buisness. 27 Jan, Berlin
87. Coutts TJ, Wanlass MW, Ward JS, Johnson S (1996) A review of recent advances in thermophotovoltaics. Proceeding of the 25th IEEE photovoltaic specialists conference, pp 25–30
88. Nelson R (2003) TPV Systems and state-of-the-art development. Proceeding of the 5th conference on thermophotovoltaic generation of electricity, Rome, 16–19 Sept 2002. American Institute of Physics, pp 3–3
89. Weinberg F (1986) Advanced combustion methods. Academic Press, London
90. Fraas L, Minkin L (2007) TPV History from 1990 to Present and Future Trends. Proceeding of the 7th world conference on thermophotovoltaic generation of electricity, Madrid, 25–27 Sept 2006. American Institute of Physics, pp 3–23
91. DeBellis CL, Scotto MV, Fraas L, Samaras J, Watson RC, Scoles SW (1999) Component development for 500 watt diesel fueled portable thermophotovoltaic (TPV) power supply. Proceeding of the 4th NREL conference on thermophotovoltaic generation of electricity, Denver, Colorado, 11–14 Oct 1998. American Institute of Physics, pp 362–370
92. Stöcker H (2000) Taschenbuch der Physik (in German), 4th edn. Verlag Harri, Deutsch

93. Broman L, Marks J (1995) Co-generation of electricity and heat from combustion of wood powder utilizing thermophotovoltaic conversion. Proceeding of the 1st NREL conference on thermophotovoltaic generation of electricity, Copper Mountain, Colorado, 24–28 July 1994. American Institute of Physics, pp 133–138
94. Osborn PD (1985) Handbook of energy data and calculations: Including directory of products and services. Butterworth-Heinemann, London
95. Schroeder KL, Rose MF, Burkhalter JE (1995) An experimental investigation of hybrid kerosene burner configurations for TPV applications. Proceeding of the 1st NREL conference on thermophotovoltaic generation of electricity, Copper Mountain, Colorado, 24–28 July 1994. American Institute of Physics, pp 106–118
96. Liley PE, Thomson GH, Friend DG, Daubert TE, Buck EB (1997) Section 2: Physical and chemical data. In: Perry RH (ed) Perry's chemical engineers' handbook. McGraw-Hill, New York
97. Adair PL, Rose MF (1995) Composite emitters for TPV systems. Proceeding of the 1st NREL conference on thermophotovoltaic generation of electricity, Copper Mountain, Colorado, 24–28 July 1994. American Institute of Physics, pp 245–262
98. Siegel R, Howell J (2001) Thermal radiation heat transfer, 4th edn. Taylor and Francis, London
99. Baukal CE (2000) Heat transfer from burners. In: Baukal CE (ed) Heat transfer in industrial combustion, Chap 8. CRC Press, Boca Raton, pp 285–343
100. Gaydon AG (1974) The spectroscopy of flames, 2nd edn. Chapman and Hall, London
101. Nelson RE (1995) Thermophotovoltaic emitter development. Proceeding of the 1st NREL conference on thermophotovoltaic generation of electricity, Copper Mountain, Colorado, 24–28 July 1994. American Institute of Physics, pp 80–96
102. Nelson RE (1992) Fibrous emissive burners Selective and Broadband. Annual Report, Gas Research Inst., GRI-92/08
103. Bitnar B, Durisch W, Mayor J-C, Sigg H, Tschudi HR (2002) Characterisation of rare earth selective emitters for thermophotovoltaic applications. Sol Energy Mater Sol Cells 73:221–234
104. Palfinger G (2006) Low dimensional Si/SiGe structures deposited by UHV-CVD for thermophotovoltaics. Doctoral thesis, Paul Scherrer Institute
105. Wünning JG (2003) FLOX–flameless combustion, Thermprocess Symposium, Düsseldorf, Verband Deutscher Maschinen- und Anlagenbau e. V. (VDMA)
106. Dryden IGC (1975) The efficient use of energy. IPC Science and Technology Press, Guildford
107. Mattarolo G (2007) Development and modelling of a thermophotovoltaic system, Doctoral thesis, University of Kassel
108. Carlson RS, Fraas LM (2007) Adapting TPV for use in a standard home heating furnace. Proceeding of the 7th World conference on thermophotovoltaic generation of electricity, Madrid, 25–27 Sept 2006. American Institute of Physics, pp 273–279
109. Qiu K, Hayden A (2004) A novel integrated TPV power generation system based on a cascaded radiant burner. Proceeding of the 6th International conference on Thermophotovoltaic Generation of Electricity, Freiburg, Germany, 14–16 June 2004. American Institute of Physics, pp 105–113
110. Fraas L, Minkin L, She H, Avery J, Howells C (2004) TPV power source using infrared-sensitive cells with commercially available radiant tube burner. Proceeding of the 6th international conference on thermophotovoltaic generation of electricity, Freiburg, Germany, 14–16 June 2004. American Institute of Physics, pp 52–60
111. Becker FE, Doyle EF, Shukla K (1999) Operating experience of a portable thermophotovoltaic power supply. Proceeding of the 4th NREL conference on Thermophotovoltaic generation of electricity, Denver, Colorado, 11–14 Oct 1998. American Institute of Physics, pp 394–412
112. Rumyantsev VD, Khvostikov VP, Sorokina O, Vasilev AI, Andreev VM (1999) Portable TPV generator based on metallic emitter and 1.5-amp GaSb cells. Proceeding of the 4th

NREL conference on thermophotovoltaic generation of electricity, Denver, Colorado, 11–14 Oct 1998. American Institute of Physics, pp 384–393
113. Doyle E, Shukla K, Metcalfe C (2001) Development and demonstration of a 25 Watt thermophotovoltaic power source for a hybrid power system. Report, NASA, TR04-2001
114. Basu S, Chen Y-B, Zhang ZM (2007) Microscale radiation in thermophotovoltaic devices—A review. Int J Energy Res 31(6–7):689–716
115. Nelson RE (2003) A brief history of thermophotovoltaic development. Semicond Sci Technol 18:141–143
116. Energy sources and systems (1997) In: Energy-efficient technologies for the dismounted soldier, Chap 3. National Academy Press
117. Hess HL (2002) Front end analysis of mobile electric power research and development for the 2015–2025 time frame, Report, Fort Monmouth, US [Online]. Available at: http://www.ee.uidaho.edu/ee/power/hhess/FrontEndAnalysis.pdf. Accessed 21 Aug 2004
118. Kruger JS (1997) Review of a workshop on thermophotovoltaics organized for the army research office. Proceeding of the 3rd NREL conference on thermophotovoltaic generation of electricity, Denver, Colorado, 18–21 May 1997. American Institute of Physics, pp 23–30
119. Aubyn J, Platts J, Aubyn JD (1992) Uninterruptible power supplies, IEE Power Series 14, Peter Peregrinus
120. Brownlie GD (1998) Battery requirements for uninterruptible power-supply applications. J Power Sources 23(1–3):211–220
121. Baker AM (2002) Future power systems for space exploration—executive summary, Report, QinetiQ UK [Online] http://esamultimedia.esa.int/docs/gsp/completed/comp_sc_00_S54.pdf. Accessed 28 April 2010
122. Milton BE (1995) Thermodynamics, combustion and engines. Stanley Thornes Pub Ltd, England
123. West EM, Connelly WR (1999) Integrated development and testing of multi-kilowatt TPV generator systems. Proceeding of the 4th NREL conference on thermophotovoltaic generation of electricity, Denver, Colorado, 11–14 Oct. 1998. American Institute of Physics, pp 446–456
124. Morrison O, Seal M, West E, Connelly W (1999) Use of a thermophotovoltaic generator in a hybrid electric vehicle. Proceeding of the 4th NREL conference on thermophotovoltaic generation of electricity, Denver, Colorado, 11–14 Oct 1998. American Institute of Physics, pp 488–496
125. Mazzer M, de Risi A, Laforgia D, Barnham K, Rohr C (2000) High Efficiency Thermophotovoltaics for Automotive Applications. Proceeding of the world congress of the engineerung society for advanced mobility land sea air and space (SAE), Detroit, Michigan, 6–9 Mar 2000-01-0991
126. Vázquez J, Sanz-Bobi MA, Palacios R, Arenas A (2002) State of the Art of Thermoelectric Generators Based on Heat Recovered from the Exhaust Gases of Automobiles. Proceeding of the 7th European workshop on thermoelectrics, Pamplona, Spain, Oct. 2002, Paper 3
127. Erickson TA, Lindler KW, Harper MJ (1997) Design and construction of a thermophotovoltaic generator using turbine combustion gas. Proceeding of the 32nd intersociety energy conversion engineering conference, IEEE, pp 1101–1106
128. Tompsett GA, Finnerty C, Kendall K, Alston T, Sammes NM (2000) Novel applications for micro-SOFCs. J Power Sources 86:376–382
129. Nelson R, Doyle E, Hurley J (1997) Utility thermophotovoltaic cogeneration, Final Report, Gas Research Institute, US, GRI-97/0416
130. Kassakian JG (2000) Automotive electrical systems–the power electronics market of the future. Applied power electronics conference and exposition, New Orleans, 6–10 Feb 2000
131. Kassakian J (2006) Team plugs into fuel-efficiency. MIT Tech Talk 50(27):4–6
132. Caton JA, Turner WD (1996) Cogeneration. In: Kreith F, West RE (eds) CRC handbook of energy efficiency, Chap 3. CRC Press, Boca Raton, pp 669–711

133. Combined heat and power (1994) In: An appraisal of UK Energy Research, Development, Demonstration & Dissemination. ETSUR83, Her Majestys Stationery Office (HMSO), pp 465–480
134. Micro–map mini and micro CHP, market assessment and development plan (2002) Summary Report, A study supported by theEuropean commission SAVE programme, DGTREN, FaberMaunsell Ltd, UK
135. Iles P, Hindman D (1995) Workshop 4: converter cooling & recuperation. Proceeding of the 1st NREL conference on thermophotovoltaic generation of electricity, Copper Mountain, Colorado, 24–28. July 1994, American Institute of Physics, pp 3–22
136. Dincer I (2002) Thermal energy storage systems as a key technology in energy conservation. Int J Energy Res 26:567–588
137. Graham G, Cruden A, Hart J (2002) Assessment of the implementation issues for fuel cells in domestic and small scale stationary power generation and CHP applications, Report F/03/00235/REP, DTI/Pub URN 02/15, Department of Trade and Industry (DTI), UK, Energy Technology Support Unit (ETSU), Harwell UK
138. Crozier-Cole T, Jones G (2002) The potential market for micro CHP in the UK, Report to the Energy Saving Trust (EST), Edinburgh [Online] Available at:http://www.energysavingtrust.org.uk/uploads/documents/aboutest/Potential_market_for_micro_CHP_2002.pdf. Accessed 14 May2011
139. Fraas L, Ballantyne R, She-Hui, Shi-Zhong Ye, Gregory S, Keyes J, Avery J, Lamson D, Daniels B (1999) Commercial GaSb cell and circuit development for the Midnight Sun(R) TPV stove. Proceeding of the 4th NREL conference on thermophotovoltaic generation of electricity, Denver, Colorado, 11–14 Oct 1998. American Institute of Physics, pp 480–487
140. Fraas LM, Avery JE, Daniels WE, Huang HX, Malfa E, Testi G (2001) TPV tube generators for apartment building and industrial furnace applications. Proceeding of the 3rd European photovoltaic solar energy conference, Munich, 22–26 Oct 2001. WIP, p 2304
141. Kushch AS, Skinner SM, Brennan R, Sarmiento PA (1997) Development of a cogenerating thermophotovoltaic powered combination hot water heater/hydronic boiler. Proceeding of the 3rd NREL conference on thermophotovoltaic generation of electricity, Denver, Colorado, 18–21 May 1997. American Institute of Physics, pp 373–386
142. Aicher T, Kästner P, Gopinath A, Gombert A, Bett AW, Schlegl T, Hebling C, Luther J (2004) Development of a novel TPV power generator. Proceeding of the 6th international conference on thermophotovoltaic generation of electricity, Freiburg, Germany, 14–16 June 2004. American Institute of Physics, pp 71–78
143. Mattarolo G, Bard J, Schmid J (2004) TPV-Application as small back-up generator for standalone photovoltaic systems. Proceeding of the 6th international conference on thermophotovoltaic generation of electricity, Freiburg, Germany, 14–16 June 2004. American Institute of Physics, pp 133–141
144. Schubnell M, Gabler H, Broman L (1997) Overview of European activities in thermophotovoltaics. Proceeding of the 3rd NREL conference on thermophotovoltaic generation of electricity. Denver, Colorado, 18–21 May 1997. American Institute of Physics, pp 3–22
145. The micro-chp technologies roadmap—Meeting 21st century residential energy needs (2003) Report, United States Department ofEnergy, Office of Energy Efficiency and Renewable Energy, Distributed Energy Program, US
146. Brown M (2009) Small-scale cogeneration in Europe: Technology status and market development, Presentation, Delta Energy and Environment, UK, Cogen España National Conference, Barcelona, 3 Nov
147. Pehnt M, Praetorius B, Schumacher K, Fischer C, Schneider L, Cames M, Voß J-P (2004) Micro CHP–a sustainable innovation, Transformation and innovation in power systems (TIPS). Socio-ecological Research Framework, Germany
148. Durisch W, Grob B, Mayor JC, Panitz JC, Rosselet A (1999) Interfacing a small thermophotovoltaic generator to the grid, 4. NREL conference on thermophotovoltaic

generation of electricity, Denver, Colorado, 11–14 Oct 1998. American Institute of Physics, pp 13–414
149. Future Practice Report No. 32 (1993) A Technical and Economic Assessment of Small Stirling Engines for Combined Heat and Power, Energy Technology Support Unit (ETSU), Harwell UK
150. Good practice Guide 3 (1995) Introduction to small—scale combined heat and power, Energy Technology Support Unit (ETSU), Harwell, UK
151. Durisch W, Bitnar B, Roth F, Palfinger G (2003) Small thermophotovoltaic prototype systems. Sol Energy 75:11–15
152. Keyes, JB (2001) Glass project fact sheet: Thermophotovoltaic electric power generation using exhaust heat, US. Department of Energy, Office of Industrial Technologies [Online] Available at: http://www.oit.doe.gov/inventions. Accessed 25 Sept 2001
153. Goldstein M.K, DeShazer LG, Kushch AS, Skinner SM (1997) Superemissive light pipe for TPV applications. In: Proceeding of the 3rd NREL conference on thermophotovoltaic generation of electricity, Denver, Colorado, 18–21 May 1997. American Institute of Physics, pp 315–326
154. DeShazer LG, Kushch AS, Chen KC (2001) Hot Dielectrics as Light Sources for TPV Devices and Lasers, NASA Tech Briefs, Photonics Tech Briefs, June [Online] Available at: http://www.techbriefs.com. Accessed 5 May 2011
155. Goldstein MK (1996) Superemissive light pipes and photovoltaic systems including same, Quantum Group Inc., US Patent 5500054 see http://www.freepatentsonline.com/5500054.html
156. Bauer T (2005) Thermophotovoltaic applications in the UK: Critical aspects of system design, PhD thesis, Northumbria University, UK
157. Theiss TJ, Conklin JC, Thomas JF, Armstrong TR (2000) Comparison of prime movers suitable for USMC expeditionary power sources, report, Oak Ridge National Laboratory, US, ORNL/TM-2000/116
158. Coutts TJ (1999) A review of progress in thermophotovoltaic generation of electricity. Renew Sustain Energy Rev 3(2–3):77–184
159. Bauer T, Forbes I, Pearsall N (2004) The potential of thermophotovoltaic heat recovery for the UK industry. Int J Ambient Energy 25(1):19–25
160. Brown HL (1996) Energy Analysis of 108 Industrial Processes. Fairmont Press, New York
161. de Beer J, Worrell E, Blok K (1998) Future technologies for energy-efficient iron and steel making. Ann Rev Energy Environ 23:123–205
162. An appraisal of UK energy research, development, demonstration and dissemination (1994) Energy Technology Support Unit(ETSU), Report R83, vol 2, London: Her Majestys Stationery Office (HMSO)
163. Good Practice Guide 13 (1999) Waste heat recovery from high temperature gas streams, Energy Technology Support Unit (ETSU), Harwell, UK
164. Rowe DM, Gao M (1998) Evaluation of thermoelectric modules for power generation. J Power Sources 73:193–198
165. Korobitsyn M (2002) Industrial applications of the air bottoming cycle. Energy Convers Manag 43:1311–1322
166. Larjola J (1995) Electricity from industrial waste heat using high-speed organic Rankine cycle (ORC). Int J Prod Econom 41:227–235
167. Product PowerUnit^{TM} (2010), Stirling Biopower, US [Online] Available at: http://www.stirlingbiopower.com/. Accessed 5May 2010
168. Wu C (1996) Analysis of waste heat thermoelectric power generators. Appl Therm Eng 16:63–69
169. Lambrecht A, Böttner H, Nurnus J (2004) Thermoelectric energy conversion—overview of a TPV alternative. Proceeding of the 6th international conference on thermophotovoltaic generation of electricity, Freiburg, Germany, 14–16 June 2004. American Institute of Physics, pp 24–32

Index

Lightning Source UK Ltd.
Milton Keynes UK
UKOW05n1624290514

232530UK00001B/32/P

9 783642 199646